Peer Review and Manuscript Management in Scientific Journals

Peer Review and Manuscript Management in Scientific Journals

Guidelines for Good Practice

By

Irene Hames

Published in association with the Association of Learned and
Professional Society Publishers (www.alpsp.org)

BLACKWELL PUBLISHING
350 Main Street, Malden, MA 02148-5020, USA
9600 Garsington Road, Oxford OX4 2DQ, UK
550 Swanston Street, Carlton, Victoria 3053, Australia

First published 2007 by Blackwell Publishing Ltd

1   2007

Library of Congress Cataloging-in-Publication Data

Hames, Irene.
    Peer review and manuscript management in scientific journals : guidelines
for good practice / by Irene Hames.
        p. cm.
    Includes bibliographical references and index.
    ISBN-13: 978-1-4051-3159-9 (pbk. : alk. paper)
    ISBN-10: 1-4051-3159-4 (pbk. : alk. paper) 1. Scientific literature—Evaluation.
2. Science—Periodicals—Evaluation. 3. Technical writing—Evaluation—Analysis.
4. Peer review. 5. Science—United States—Information services. I. Title.

Q225.5.H255 2007
070.4′495—dc22

                                                    2006032606

A catalogue record for this title is available from the British Library.

Set in 10/13pt Rotis Semisans
by Graphicraft Limited, Hong Kong
Printed and bound in the United Kingdom
by TJ International, Padstow, Cornwall

The publisher's policy is to use permanent paper from mills that operate
a sustainable forestry policy, and which has been manufactured from
pulp processed using acid-free and elementary chlorine-free practices.
Furthermore, the publisher ensures that the text paper and cover board used
have met acceptable environmental accreditation standards.

For further information on
Blackwell Publishing, visit our website:
www.blackwellpublishing.com

Contents

Foreword

Henry Oldenburg, the founder of the first English-language research journal (*Philosophical Transactions of the Royal Society*), understood very clearly the importance of peer review – indeed this task was carried out for him by the Council of the Royal Society (not a task any longer expected of society councils!). Despite the challenges to many aspects of journal publication, and despite criticisms of peer review and experiments with alternative approaches, survey after survey shows that authors and readers still place great value on the filtering and improvement provided by the peer-review process, and wish to preserve it regardless of what else may change.

These days, around 1.8 million peer-reviewed articles are published every year; if each is reviewed by at least two reviewers, this means that at least 3.6 million reviewers' reports are produced every year (probably more, as some articles are resubmitted to several journals before finally being accepted). Given the scale of the operation, it is surprising that no one has previously attempted to write a handbook on how the process should be carried out.

The need for this book is obvious – ALPSP is frequently asked where 'the rules' for proper peer review can be found, and until now there has been no satisfactory answer. Fortunately for journal editors and publishers, Irene Hames is ideally qualified for the task. She set up the editorial office of *The Plant Journal* in 1990 and the first issue was published in July 1991; in 2006 it published over 4000 pages, achieved an Impact Factor of 6.97 and worked with over 1000 reviewers. Irene is a knowledgeable and popular speaker at journal publishing meetings, and has also found time to advise others on how to run an editorial office. Her experience of managing the complex processes of peer review and manuscript management is apparent in this highly practical book, which will undoubtedly become the 'bible' of peer review, not only for those working in the sciences but also for those in the arts and humanities. No editor or publisher should be without it.

Robert Campbell, Blackwell Publishing
Sally Morris, Association of Learned and
Professional Society Publishers

Preface

Editors are frequently called the 'gatekeepers' of their journals. They are most certainly this, determining what their journals will publish and what they won't. But their role goes beyond this. They also act as 'midwives', bringing to fruition the labours of researchers – highlighting experimental inadequacies, pointing out misinterpretations of results and offering alternative explanations, improving the presentation of manuscripts, and advising on alternative and perhaps more appropriate venues for the publication of their work. Editors are also the guardians of the scholarly record, with a duty to ensure that this is kept free from corrupting influences and that errors – both genuine and those resulting from fraudulent work or unethical behaviour – are corrected appropriately and as quickly as possible. Editors are given great power when they are appointed, and with this comes enormous responsibility. Suddenly they are entrusted with the work of other researchers, perhaps even that of their competitors, and their decisions determine whether or not that work is published. Because success in job and grant applications, career advancement and public recognition depend to a large degree on publication records, they are therefore indirectly responsible for these.

At the heart of all these roles lies the peer-review process, nowadays oft-maligned, but at its best a very powerful and sophisticated tool. Yet in the great majority of cases, editors come to the job without any specific knowledge about or training in peer review, the decision-making process, or the potential problems and pitfalls. For many it has been difficult to know where to find advice and guidance. My aim in writing this book has been to provide a manual to help editors and their editorial colleagues and staff – both those new to their roles and those who have been in post some time but may be struggling or unhappy with their procedures. I wanted first and foremost to provide practical guidance on all aspects of peer review and create an awareness of the issues involved and the potential problems.

Editors and editorial staff are amongst the most committed and enthusiastic of individuals, frequently becoming passionate about their journals. I have been fortunate in my years with *The Plant Journal* to experience this at first hand, and my sincere thanks go to all the editors and editorial office staff with whom I have had the privilege of working over the past 16 years. My thanks go also to those individuals who provided me with invaluable feedback and constructive comments during the writing of this book – Sally Morris from ALPSP, Bob Campbell and Edward Wates from Blackwell Publishing, Alex Williamson from the *BMJ* Publishing Group, and three anonymous reviewers. The last were clearly people with great experience of journal editorial work and so their positive reactions and comments were a source of great encouragement and help. But the final responsibility for the content lies with me and I hope that even if there are things with which some readers may disagree, it will make them think about the problems and perhaps reassess their own procedures. The book is not intended to be prescriptive but rather a source of guidance and help. All journals (and the communities they serve) are different, and their editors are the people who are best placed to decide what is most appropriate for them. The book has been four years in the writing and my thanks go to Erica Schwarz for so ably steering it through the production process to become a reality.

Irene Hames

1 Introduction

The practice of reviewing manuscripts for publication has been around for nearly 300 years, since the Royal Societies of Edinburgh* and London started seeking the advice of their members in the early to mid-18th century to help them select articles for publication.[1] Gradually, a number of other scientific and professional societies adopted the practice, but procedures developed in a rather haphazard and ad hoc way. Peer review, the process by which material submitted for publication is critically assessed by external experts (see Box 1.1), was introduced into different journals at different times and in different ways, often dependent on the chief editor at the time. It is only since the middle of the 20th century that it has become generally widespread and reasonably standardized. Excellent accounts of the origins and evolution of editorial peer review can be found in the articles by Kronick[1] and Burnham,[2] respectively.

Two main factors led to the spread of peer review. Firstly, until the relatively recent past, editors frequently had to struggle to find enough material to publish and so did not need to be selective. Over the past 50 years this has changed, to the point where submissions to scientific journals are burgeoning and editors need to be highly selective in what they publish in their journals. Secondly, as scientific areas expanded and became increasingly specialized and sophisticated, editors were no longer able to be experts in all areas. They needed to seek the opinion and advice of others. Today, peer review is used almost universally by scientific journals, and a peer-reviewed journal is generally considered to be 'one that submits most of its published research articles for outside review', i.e. by 'experts who are not part of the editorial staff' (as defined by the International Committee of Medical Journal Editors, ICMJE[3]).

Scholarly publication is the means by which new work is communicated, and peer review is a vitally important part of the publication process. It is the quality-control

---

Box 1.1 Definition of peer review

'Peer review is the critical assessment of manuscripts submitted to journals by experts who are not part of the editorial staff.'

International Committee of Medical Journal Editors (ICMJE). Uniform Requirements for Manuscripts Submitted to Biomedical Journals.[3]

---

* The Royal Society of Edinburgh was created in 1783 from its forerunner, the Philosophical Society of Edinburgh. This was originally founded in 1731 as the Society for the Improvement of Medical Knowledge but changed its name in 1737 to reflect broadening interests.

mechanism that determines what is and what is not published, and in most scientific disciplines work will not be considered seriously until it has been validated by peer review. It also acts as a filter for interest and relevance. Publication is of central importance in both academic promotion and the allocation of research funds. It is the means by which scientific discoveries are attributed to individuals. In some areas, this establishment of priority can lead to very significant commercial and financial rewards. Since so much hinges on peer review and it is so central to what and where things are published, it is essential that it is carried out well and professionally, and that it is viewed with confidence and respect. There has, however, been a growing movement, particularly in biomedical publishing, to highlight its shortcomings.[4,5] Critics of peer review cite examples that point to its failure, because of its conservatism, to recognize important and innovative papers; its failure to spot errors; its lack of consistency and objectivity; its poor record in detecting fraud; its openness to abuse and bias; and to it being labour intensive, expensive, and often slow, with resulting delays in publication. These critics of peer review suggest there is little evidence to support the use of peer review as a mechanism to assure the quality of research publications, and frequently state that it is only the lack of an obvious alternative that keeps the process going. There have been calls for funding for large-scale research programmes to look into the effectiveness of peer review and potential alternatives.[6]

Despite all the criticisms and reputed failings of peer review, it is inescapable that it is very extensively used in scholarly publishing. Many editors are, in fact, very pro-peer review and would agree with Laine and Mulrow,[7] who have stated (page 1038), 'We cannot imagine getting along without peer review', and who 'salute' the individuals who review for them. Five surveys carried out between 1999 and 2005 have confirmed the importance with which peer review is viewed and the widespread feeling that the accuracy and quality of material that has not been peer reviewed cannot be trusted.[8-12] The surveys have also, however, brought to light considerable dissatisfaction with reviewing standards and the peer-review process, especially regarding its quality and fairness, and about the delays that can occur.

Peer review is, therefore, extremely important and is likely to be around for quite some time. Various modifications have been suggested and new systems are being tested, but 'traditional' peer review remains the method practised by the great majority of scientific journals. There are clearly, however, concerns about the quality and speed of peer review and there is, therefore, scope for improvement.

What should peer review do?

What should peer review do? Ideally, it should:

- prevent the publication of bad work – filter out studies that have been poorly conceived, designed or executed

- check that the research reported has been carried out well and there are no flaws in the design or methodology
- ensure that the work is reported correctly and unambiguously, with acknowledgement to the existing body of work
- ensure that the results presented have been interpreted correctly and all possible interpretations considered
- ensure that the results are not too preliminary or too speculative, but at the same time not block innovative new research and theories
- select work that will be of the greatest interest to the readership
- provide editors with evidence to make judgements as to whether articles meet the selection criteria for their particular publications
- generally improve the quality and readability of a publication (although this is more a by-product of peer review).

So, fundamentally, peer review maintains standards and ensures reporting is as truthful and accurate as possible. It helps the layperson or non-expert assess what to believe and what to view with scepticism. With the advent of the World Wide Web, arguments abound that everything should be published and be available to everyone for them to make their own evaluations. But how can non-specialists evaluate and make judgements about things they know nothing about? It is difficult enough for scientists outside of their fields of expertise to assess the merits of competing claims, and so almost impossible for the layperson. This has led to the argument that what is needed is more, not less, quality control and the involvement of the best and most expert individuals to ensure there is genuine review by peers.[13]

The peer-review process needs to be handled efficiently and effectively. It must help journals provide the type and quality of material they are aiming to publish for their specific audiences. Reviewers need therefore to understand the quality and scope of the journals they review for. They need to be provided with guidelines on this. Authors need to be 'trained' to recognize the scope and standard of paper that is required for a particular journal. Editors need to select the most appropriate reviewers, taking care not to overload them. Editors are responsible for ensuring the quality of their journals and that what is reported is ethical, accurate and relevant to their readership (see Golden Rule 1).

Golden Rule 1

Editors are responsible for ensuring the quality of their journals and that what is reported is ethical, accurate and relevant to their readership.

What does peer review assume?

The peer-review process depends on trust and requires the goodwill and good behaviour of all the participants, i.e. the authors, reviewers and editor. It assumes certain things. It assumes that authors are submitting original work that has been honestly carried out, evaluated and reported. Journals cannot be expected to detect fraud at the laboratory experimental level – that is not their role. It assumes that reviewers assess submitted papers to the best of their ability in a courteous and expeditious manner, respecting the confidentiality of submitted material and disclosing any potential conflict of interest. And it assumes that editors evaluate all the information available to them and make decisions on whether to publish material or not as fairly and transparently as possible. It is important to remember that it is not the reviewers who decide what will or will not be published. They assess and advise, commenting on quality and suitability and alerting editors to flaws and problems, but it is the responsibility of editors to decide what will be published in their journals. In making these decisions, it may be helpful for them to bear in mind something very wise that Stephen Lock, a former editor of the *BMJ* (*British Medical Journal*), wrote in his seminal book on peer review, 'A Difficult Balance', in 1985 (page 129)[14]:

> Peer review does not, and cannot, ensure perfection: scientific journals are records of work done and not of revealed truth. If they were to insist on absurdly high standards science would suffer more than it would gain, purchasing reliability at the expense of innovative quality.

What is this book trying to achieve?

Given that peer review is used by the vast majority of scientific journals, the fundamental role it plays in scholarly publishing and the great importance with which it is viewed, it should be carried out to the highest possible standards. Peer review is a very powerful tool if used correctly, but as in every area of life, the whole spectrum of quality exists, from very poor to excellent. It is also rather an 'amateur' activity in that there is usually no formal training, with most people learning 'on the job'. My aim is therefore to provide guidelines for good practice that will be useful to journals of all sizes, in many scientific disciplines. Although the book is primarily for people in science, there is much that will be applicable to other scholarly areas, as the general principles and many of the procedures are the same.

Editorial offices range from organizations where one person does everything to those where many people are employed. Some journals, particularly large ones, have central offices that remain through changes of editors and in which a large body of expertise has been built up over time. They are frequently overseen by a managing editor or equivalent. For others, the office moves every time there is a change in editor-in-chief. As editors are usually appointed on the basis of their academic standing and expertise, and frequently for their visionary aims and aspirations for their journals, it is highly likely they may have no direct practical experience of running a peer-review system, and certainly may never have had to set one up from scratch. It is not unusual for assistants appointed to have no or very little idea what to do. Yet, the filing cabinets, computers, and so on may arrive one day at a new location, and the office will be expected to be up and running the next.

This is a basic 'how to' guide for people involved in editorial peer review – journal editors, editorial office staff and publishers; a handbook that can be dipped into as required or read in entirety without too much effort. My hope is that the contents of this book will be of help to the newcomer to peer review, as well as acting as a refresher and useful reference for those with experience but who may have gaps in their knowledge or want to review their current practices. All the practical aspects of peer review are covered: from how to set up and run an efficient peer-review system to dealing with unusual and sensitive situations, from manuscript submission to final decision. Scientific review and publication can get caught up in political, ethical or moral questions. I hope the book will provide help to editors and editorial office staff to make things more straightforward and reduce the impact such issues might have on peer review and decisions on whether or not to publish.

I wanted to avoid swamping readers with references, especially as most editors and editorial office staff are very short of time, so only those that are useful or important are given. Two books have been published which will be of interest to readers wanting to find out more about specific aspects of peer review and the research that has been done and is going on. In her book on editorial peer review, Ann Weller reports the results of a systematic review of published studies on the editorial peer-review process, covering all English language studies published between 1945 and 1997 (she was not able to locate any studies published before 1945).[15] The book on peer review edited by Fiona Godlee and Tom Jefferson goes into many aspects of peer review in the health sciences and is a great source of information.[16] It is also very readable. There have been five International Congresses held on peer review since 1989, where research, rather than opinion, relevant to peer review has been presented. The Congresses were initiated by the *Journal of the American Medical Association* (*JAMA*) following an article by Bailar and Patterson in 1985,[17] in an attempt to bring the rigours of scientific enquiry to peer review. Papers presented at the first four Congresses appeared in *JAMA*.[18–21] Summaries of the presentations given at the latest Congress, held in 2005, can be found on the Congress website.[22] Papers based on the presentations will be published in a number of places.

Chapter 2 starts with the basics – it describes the peer-review process and how to go about setting up a peer-review system from scratch, or how to improve an existing one. It gives some thought to the people involved as well as to the systems and procedures that are needed. The third chapter deals with the first stage of the peer-review process, a pre-review stage really, but crucial to achieving a successful and thorough review – manuscript submission, and the checks and evaluations that need to be carried out to ensure manuscripts are complete and suitable for a journal. If they are and they make it past the initial assessment, they will go into the full review process. This is dealt with in Chapter 4, which covers the whole process: identifying, selecting and contacting reviewers; sending them manuscripts and all other necessary material; monitoring the review process and chasing up reviewers; receiving and checking reviews. This chapter describes the checks that need to be made at each stage and the sorts of problems, or unusual situations, that can arise and how to deal with them. Answering enquiries from authors (and others) is also covered, as this is not always as simple as it seems and there are pitfalls to be avoided. In Chapter 5 we move on to the decision-making step. The organizational structure for this will vary from journal to journal, and will depend partly on journal size and complexity and partly on practical considerations. The various possibilities are discussed, along with the range of editorial decisions that can be made and the factors to take into account when making decisions, including dealing with dual-use research. Communicating decisions to authors is covered and there is consideration of revisions, resubmissions and the final acceptance stage, along with rebuttals and appeals.

Good practice in peer review is system and business-model independent, so the guidelines given throughout this book apply to both paper-based and online systems, and to both subscription and author-side-payment business models. Special considerations that are relevant to either paper or online are given whenever appropriate. Online submission and review is an important and relatively new area. Moving to online working and making a successful transition are covered in Chapter 6. This includes information on how to go about choosing an online system, how to prepare for the move to online working and how to implement a new system. It also describes what journals can expect after the move and the problems that may be encountered, with suggestions on how to deal with them.

Peer review could not survive without reviewers – they are truly a precious resource, and Chapter 7 gives guidance on how they should be treated and offers some suggestions on ways to compensate them for the time and effort they give to journals. Authors, reviewers and editors all have obligations and responsibilities, and ethical standards to which they should adhere. These are described in Chapter 8, along with conflicts of interest and certain moral dilemmas editors may find themselves facing. Chapter 9 covers the various forms of misconduct in scientific research and publishing, many of which, unfortunately, seem to be on the increase. It includes advice on how to handle cases of alleged or suspected misconduct, and on where editors and journals can turn for help. Measures that need to be taken to correct the literature are also described.

I've drawn up a list of 14 basic principles for peer review and called these 'the Golden Rules'. As well as being numbered and highlighted in the text (the first Golden Rule has already appeared, on page 3 of this chapter), they are listed in Appendix I. This appendix also contains the Peer-Review Good Practice Checklist. Here, important information from the book is summarized into Key Points and grouped under various headings. There are three more appendices: examples of various checklists, forms, guidance and editorial letters are given in Appendix II; a list and description of websites of relevance or interest appears in Appendix III; a brief description of alternative models of peer review is provided in Appendix IV, along with details of where to go to find out more about them. Readers are also alerted throughout the book to things that may be problematic or where they should be cautious; these are labelled 'Beware!' and appear in boxes.

As already mentioned, peer review can be a very powerful tool. It is hoped that the guidelines in this book will help editors and others achieve the highest standards of reviewing practice, to the benefit of both their own journals and scholarly publishing in general.

References

1   Kronick, D. A. (1990). Peer review in 18th-century scientific journalism. *JAMA*, **263**, 1321–1322.
2   Burnham, J. C. (1990). The evolution of editorial peer review. *JAMA*, **263**, 1323–1329.
3   International Committee of Medical Journal Editors. Uniform Requirements for Manuscripts Submitted to Biomedical Journals: Writing and Editing for Biomedical Publication. www.icmje.org (updated February 2006; accessed 25 May 2006).
4   Wager, E. and Jefferson, T. (2001). Shortcomings of peer review in biomedical journals. *Learned Publishing*, **14**, 257–263.
5   Smith, R. (2006). Peer review: a flawed process at the heart of science and journals. *Journal of the Royal Society of Medicine*, **99**, 178–182.
6   Jefferson, T. O., Alderson, P., Davidoff, F. and Wager, E. (2003). Editorial peer-review for improving the quality of reports of biomedical studies. *The Cochrane Database of Methodology Reviews* 2001, Issue 3. Art. No.: MR000016. DOI: 10.1002/14651858.MR000016.
7   Laine, C. and Mulrow, C. (2003). Peer review: Integral to science and indispensable to *Annals. Annals of Internal Medicine*, **139**, 1038–1040.
8   Key Perspectives Ltd. (1999). What Authors Want: The ALPSP research study on the motivations and concerns of contributors to learned journals. ALPSP. www.alpsp.org/publications/pub1.htm.
9   ALPSP/EASE. (2000). Current Practice in Peer Review: results of a survey conducted during Oct/Nov 2000. www.alpsp.org/publications/peerev.pdf.
10 Key Perspectives Ltd. (2002). Authors and Electronic Publishing: The ALPSP research study on authors' and readers' views of electronic research communication. ALPSP. www.alpsp.org/publications/pub5.htm.
11 Key Perspectives Ltd. (2004). JISC/OSI Journal Authors Survey Report. www.jisc.ac.uk/uploaded_documents/JISCOAreport1.pdf.
12 Rowlands, I. and Nicholas, D. (2005). New Journal Publishing Models: An international survey of senior researchers. CIBER. www.ucl.ac.uk/ciber/ciber_2005_survey_final.pdf (for a summary article see Rowlands, I. and Nicholas, D. (2006). The changing scholarly communication landscape: an international survey of senior researchers. *Learned Publishing*, **19**, 31–55).
13 Gannon, F. (2001). The essential role of peer review. *EMBO Reports*, **2**, 743.
14 Lock, S. (1985). A Difficult Balance. Editorial peer review in medicine. ISI Press, Philadelphia.
15 Weller, A. C. (2001). Editorial Peer Review: Its Strengths and Weaknesses. ASIST Monograph Series, Information Today, Inc. Medford, New Jersey.
16 Godlee, F. and Jefferson, T. (Eds) (2003). Peer Review in Health Sciences (second edition). BMJ Books, London.
17 Bailar, J. C. and Patterson, K. (1985). Journal peer review: the need for a research agenda. *New England Journal of Medicine*, **312**, 654–657.
18 Guarding the guardians: research on editorial peer review. Selected proceedings from the First International Congress on Peer Review in Biomedical Publication. May 10–12, 1989, Chicago, Illinois. (1990). *JAMA*, **263**, 1317–1441.

19   The Second International Congress on Peer Review in Biomedical Publication. September 9–11, 1993, Chicago, Illinois. Selected proceedings. (1994). *JAMA*, **272**, 91–173.

20   The Third International Congress on Peer Review in Biomedical Publication. September 18–20, 1997, Prague, Czech Republic. Selected proceedings. (1998). *JAMA*, **280**, 213–302.

21   The Fourth International Congress on Peer Review in Biomedical Publication. September 14–16, 2001, Barcelona, Spain. Selected proceedings. (2002). *JAMA*, **287**, 2759–2871.

22   The Fifth International Congress on Peer Review and Biomedical Publication. September 16–18, 2005, Chicago, Illinois. Program and abstracts. (2005). www.ama-assn.org/public/peer/program.html.

2 The peer-review process – how to get going

The basic process

Whatever the size of a journal and whatever the infrastructure of its organization, the same basic steps need to be carried out in the peer-review process. Authors submit manuscripts. These need to be logged, checked to ensure they are complete and prepared according to the journal's instructions, and their receipt acknowledged. Each manuscript then needs to be read by an editor, individually or in consultation with other editors (members of the editorial board or equivalent), to assess its suitability for the journal according to guidelines determined by editorial policy. A manuscript might then be rejected without external review for one or more reasons and the author notified. If it is not rejected, it is sent out for review to external reviewers (which is a requirement of peer review; see Golden Rule 2), the number of reviewers being determined by editorial policy, most likely two in normal circumstances (see Chapter 4, page 52). Review by internal staff editors may complement this. The reviewers return their recommendations and reports to be assessed by the editor, who then makes a decision, either on his or her own or in consultation with other editors, on whether to reject the manuscript (either with or without encouragement to resubmit), to accept it pending satisfactory revision, or to accept it as it stands. For manuscripts accepted pending revision, the authors will submit a revised manuscript that will go through all or some of the above stages. Once a manuscript has been revised satisfactorily (more than one revision may or may not be allowed, again according to individual editorial policy) it can be accepted and put into the production process to be prepared for publication. This scheme at its most basic is shown in Figure 2.1. Despite the apparent simplicity of this process, the actual steps may be quite elaborate and involve a number of people and alternative procedures. Complications can arise that require problem solving or troubleshooting, and may lead to unusual steps being taken or different procedures being followed.

The basic scheme described above is system and business-model independent. However, it is very important that if a journal offers authors the option of paying to make their articles available for free access on publication, the financial aspects, including whether or not an author intends to take up the option to pay, are dealt

Golden Rule 2

Peer review must involve assessment by external reviewers.

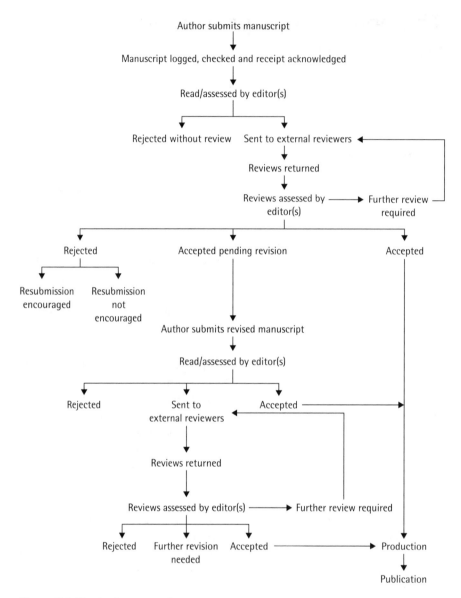

Figure 2.1 The basic peer-review process.

with after the peer-review process has been completed, or these are kept concealed from the peer-review process, so as to avoid any potential bias.

The people involved in running the peer-review process

Who does what will vary from journal to journal. In large journals with high numbers of submissions there will be a team of people involved, some of whom may have dedicated roles, for example the receiving and checking of manuscripts; sending out manuscripts to reviewers, then chasing them for their reviews and co-ordinating their return; or communicating decisions to authors. There may be a team of in-house staff editors who assess manuscripts on receipt. In small journals, all these things, and much more, will probably be done by a single individual. The titles of the people doing the various jobs may also vary from journal to journal and there is a lack of consistency in this. The title 'editor' (with the editor-in-chief or chief editor being the main editor – frequently written as 'Editor') usually refers to an academic editor, or equivalent, who will have a high level of editorial input and involvement in decision making (in this book, the term editor is used in this context). Deputy editor and assistant editor may denote someone similar but with a lower level of responsibility, or these titles may describe someone with no decision-making powers, basically someone who is running the day-to-day mechanics of the peer-review process. In some journals, someone called an editorial assistant may have greater responsibility than an 'editor' of some kind elsewhere. Titles are important, both to the individuals for their own personal satisfaction and self-esteem, and to outsiders so that it helps them understand a person's role. So when starting a new journal or setting up a new editorial office, it's worth putting some thought into what people are going to be called. If you're inheriting a journal, it's also a good idea to take a look at everyone's titles to see if they're appropriate. If you're overseeing a journal that's growing rapidly, consider carefully the titles of new people you appoint, and take the opportunity to assess those that already exist to see if they need to be changed, perhaps to reflect specific areas of responsibility or seniority that have arisen as a result of the journal's growth. Box 2.1 gives some of the common job titles found in journal editorial work. Throughout this book I will use certain titles to denote certain roles (as defined in Box 2.2) but readers should be aware that these roles may be carried out in different journals by people with different titles.

Office organization

How the peer-review process is administered will depend on whether there is one centralized editorial office managed by an editor-in-chief, managing editor or equivalent, with subject or handling editors in different geographical locations, or two or more offices in different locations, possibly dealing with submissions from different geographical areas. In the former case, all information goes through one office, which co-ordinates all activity and is aware of the status of everything that is going on and knows what everyone is doing – it is the nerve centre of the journal (or journals if the office is responsible for more than one journal, for example a group of journals belonging to a society or published by a single publisher). In the latter case, the various offices need to establish excellent communication channels and each needs to be aware of what the others are doing in order to maintain harmony and consistency.

Box 2.1 Some common titles of people involved in journal editorial work

Editor-in-chief	Managing editor
Chief editor	Editorial manager
Executive editor	Assistant editor
Deputy editor	Manuscript submissions manager
Senior editor	Manuscript submissions co-ordinator
Co-editor	Manuscript manager
Associate editor	Manuscript co-ordinator
Regional editor	Editorial co-ordinator
Section or subject editor	Editorial administrator
Receiving editor	Editorial assistant
Handling editor	
Specialist editor	
Communicating editor	
Advisory editor	
Editor	

With the arrival and increased use of web-based online manuscript submission and review systems (see Chapter 6), physical location has become much less important than it is for paper-based systems. Since all information on manuscripts, reviewers, correspondence, and so on, can be accessed via a computer with an Internet connection, editorial work can be carried out just as efficiently and effectively if individuals are thousands of miles apart as if they are in adjacent offices. Sometimes even more so – many editorial staff will have experienced the elusive local editor who is almost impossible to get hold of, or with whom any communication is difficult because they still don't use email directly and have a secretary to field calls between them and the outside world.

In any journal office, even if certain individuals have specific responsibilities, it's important for the other members of the office to know what the jobs of their colleagues involve, and to be able to provide cover when required. They need to be kept up to date with new developments and refinements to systems and procedures. Whatever the office structure, it is therefore an excellent idea, and good office practice, to develop an office manual that details all procedures and gives step-by-step instructions for each of these. Overall responsibility for the manual is best delegated to one person, but each section should be updated regularly by the person responsible for the procedures in that section. Such a manual is invaluable when cover needs to be provided for someone, to act as a refresher for those processes that are not routine, and as a guidance reference source for new staff. If there is more than one office carrying out the same procedures, it is crucial that the staff liaise to make sure these are being carried out consistently and to the same standards. If there is one main administrative office with handling editors in different locations, either

Box 2.2 Titles and roles of people involved in journal editorial work as used in this book

Title	Role
Editor-in-chief	The main or head editor. Has highest level of editorial input and decision-making powers, directs policy decisions and is responsible for the quality and content of the journal and its future direction and development.
Editor	Has high level of editorial input, is involved in formulating editorial policy with the editor-in-chief and other editors, has decision-making powers, and is accountable to the editor-in-chief.
Handling editor Subject editor Specialist editor	Various terms used to denote 'editor' in the manuscript handling and review process, where the editor is responsible for a sub-group of manuscripts determined by subject matter, geographical area of submission, or some other criterion.
Managing editor	Responsible for managing the editorial office and staff, overseeing the peer-review process, and ensuring all aspects of editorial activity run smoothly. Liaises with all parties involved in manuscript submission, handling, review, publication and promotion (authors, editorial office staff, editors, editor-in-chief, reviewers, readers and all departments at the publishers, for example production, marketing, subscriptions, rights). May or may not have decision-making powers. Oversees implementation and enforcement of editorial policy and journal development.
Editorial assistant	Assists the editor-in-chief, managing editor and editors in all areas of editorial activity from manuscript submission to publication. Interacts with authors and reviewers, providing information and assistance as required. Does not have decision-making powers. May have dedicated role within the editorial office, for example dealing with new submissions or sending manuscripts out for review.
Reviewer	Receives manuscripts for review from the journal or editor. Submits an assessment and opinion based on the quality and presentation of the work and its suitability for the journal. Provides a report for the authors and advises the journal or editor of any problems or special considerations.
Corresponding author	The co-author on a manuscript with whom a journal communicates on all matters related to that manuscript. This author is responsible for ensuring that all author guidelines are followed and that all the co-authors have approved the submission of the manuscript and have agreed to abide by all the journal's policy requirements. This author is also responsible for resolving all inter-author disputes and for dealing with all communications about the published paper.

Golden Rule 3

The submission of a manuscript and all the details associated with it must be kept confidential by the editorial office and all the people involved in the peer-review process.

Golden Rule 4

The identity of the reviewers must be kept confidential unless open peer review is used.

with or without assistants, then the main office should provide the editors and their assistants with the guidelines and operating instructions that are relevant to them, and they should be kept up to date on changes and new developments as these arise.

In each office where editorial work is being carried out, confidentiality and security issues need to be considered. It should be remembered that the *actual submission* of manuscripts is confidential, as well as the content and any information on the review of that manuscript (see Golden Rule 3). The identity of the reviewers must also be kept confidential unless open peer review is used (Golden Rule 4; and see Chapter 4, page 42). Some very simple steps can be taken to help ensure confidentiality. The location of desks needs to be thought about and how easy it would be for people passing or stopping to talk to see what is being done in any detail. Desks should be positioned to avoid, or minimize, this happening. It may be difficult to do this if editorial work is being carried out in an open-plan environment or from a corner in someone else's office, but things can still be done, as described in Box 2.3.

Choice of system and procedures

If you're setting up the peer-review system for a journal from scratch, before you do anything you need to think very carefully about how you want to work and what will be right and appropriate for your journal. Don't launch in regardless in desperation and do things in an ad hoc way. A period of thought and planning will pay off in the long run. As well as having your own ideas, it can be helpful to find out how other journals work – perhaps ones you admire or with whose editorial offices you've been impressed and had good experiences as an author, reviewer or editor. Your publisher, if you have one outside of your own organization, will also be able to advise you on this as they will have journals on their lists whose offices run efficient and much-admired review systems – ask for contact details and an introduction. Visits to other offices can be very useful, not only to find out what works well and results in a well-run office and smooth peer-review process, but also to see what to avoid doing.

Once you've decided on the structure you'd like to put in place, you might want to get feedback from your editors and other potential users. But beware, you might get

Box 2.3 Things that can help maintain confidentiality in an office environment

■ Have a screen saver that can be brought up immediately if necessary, or get into the habit of minimizing the screen as a first reaction when someone outside of the editorial team stops by your desk.

■ Do the same if you have to leave your desk or office. Log off completely if you're going to be away out of sight of your desk or for any significant time.

■ If someone else shares your computer or asks to use it, close down all editorially related programs and files. These should be password controlled so others cannot access them.

■ Don't leave paper files open or manuscripts on your desk when you're not there.

■ If someone local has submitted a manuscript, take care to keep it and any information on its review well out of sight and not easily accessible.

■ If an on-site editor has submitted a manuscript, don't file it where it can be easily come across when the editor may be looking for other manuscripts. Hide it!

■ Think carefully about what you say on the phone. If you lack the privacy needed to maintain confidentiality, try to arrange use of a phone somewhere more private for those times you need to make such calls.

■ Think carefully also about any confidential conversations you have. Close the door if you're in your office. Avoid such conversations in public places, especially at conferences where many of those attending may have direct knowledge of, or interest in, the things you're talking about.

■ If you're in a shared office, don't leave or store editorial material on surfaces or shelves. Keep them in a filing cabinet or cupboard that you can lock when you're out of the office.

completely opposing opinions, and there will always be those whose idiosyncratic requests it would be unsuitable to adopt. There may, however, be certain common elements that will reinforce your own decisions, and also some good ideas may be suggested of which you hadn't thought.

Some readers may be wondering what 'large' and 'small' are in journal terms. There is no absolute definition, and workload may depend not only on the number of manuscripts a journal has to deal with, but also on the subject areas covered by the journal and on other factors such as whether controversial or sensitive issues are involved, whether the research areas are fast moving and competitive, or whether a journal has a high profile and is a leader in its field – all these can bring problems and increase workload per manuscript and demands on time. Decision making may take longer and require the involvement of a number of people because publication may be controversial and have an impact outside of the community served by the journal, for example if public interest is high or publication may affect government economic or health policy. As a general point, systems and procedures often keep working smoothly until a certain number of submissions is reached. Then they can, quite suddenly and quickly, become inadequate or even fail. For this reason, it is crucial that all systems and procedures are regularly evaluated and staffing levels assessed to make sure they are coping well and are not about to go into overload as they approach a critical mass. Where that point is will vary from journal to journal, and will depend on the existing structure and procedures as well as on the numbers of manuscripts and individuals involved.

Record keeping

A key component of any peer-review system is its records, which need to be comprehensive, accurate and up to date. All editors and editorial office staff should make themselves aware of the current data protection and storage requirements of the country in which they are working and ensure compliance with these. In any peer-review system, various records need to be kept and a number of things need to be tracked.

Beware!

Make sure that all activity in your journal complies with the current legal data protection and storage requirements of your country of location.

Manuscripts
Manuscripts need to be tracked from the time of submission until final decision to accept or reject, and their status must be readily and quickly obtainable at any time. A system needs to be developed for regular checking of manuscripts submitted and 'active' in any way, i.e. under preliminary editorial assessment, out for external review, awaiting decision or action, being revised, or subject to a revision or resubmission deadline. No manuscript should ever get lost, be forgotten, or be unduly delayed at any stage because of oversight. Procedures for reminding people and moving manuscripts along at every stage need to be introduced. There may be need to refer back to manuscript records after editorial review has been completed, for example to answer enquiries during production, to sort out post-publication matters, and with respect to future related submissions or resubmissions. So, information needs to be kept for some time (again, subject to legal data protection and storage requirements), and policy on this needs to be developed. Information about pre-submission enquiries (see Chapter 3, page 26) and correspondence and advice from those should ideally be added to a manuscript's records.

Reviewers
A variety of names are used for the people to whom a manuscript is sent for assessment or review, the most common being reviewer, referee, assessor or advisor. It is

> **Golden Rule 5**
>
> Reviewers advise and make recommendations; editors make the decisions.

largely a matter of personal preference as to which is used, but as in all things, it is better for the term to reflect the role as much as possible. Hence, throughout this book I will use the term 'reviewer' (to reflect peer review, i.e. review by peers). The term referee, which is very commonly used, implies that the person evaluating the manuscript is an umpire, someone deciding on fair play between opposing sides and enforcing rules and settling disputes. But it is editors who have the decision-making role and who weigh up all the arguments and evidence and decide how to settle disputes, and so any names that might cause confusion in roles should be avoided. Reviewers advise and make recommendations; it is editors who make the decisions (see Golden Rule 5).

A database of potential reviewers needs to be set up (see Chapter 4, page 44). Various information on reviewers has to be obtained and regularly updated. This should include both personal information (such as contact details, areas of expertise and interest, and special requests or exclusions) and activity-related records (such as number of manuscripts reviewed, current workload, and timeliness and quality of review).

Editors

Editors' personal details relevant to editorial activity need to be kept, along with records of their current and past workloads, and which manuscripts they have handled or been involved with editorially. The status of their current manuscripts should be followed, and mechanisms put in place for reminding editors about outstanding decisions or responses. Times they will be away or not able to deal with editorial work need to be noted, so that timely and appropriate arrangements can be made to cover these periods.

Statistics and information

Every journal editor-in-chief should produce or have access to certain statistics related to their journal, for example submission numbers, geographical origin of submissions, manuscript types being submitted, acceptance and rejection rates, overall and for individual editors, reviewing times and reviewer performance, and decision and handling times, again, overall and for individuals. Examples of some of these are shown in Tables 2.1–2.3 and Figures 2.2 and 2.3. Some statistics they will want to monitor regularly, others to have available for editorial meetings. It is very important to be able to generate regular and accurate reports to help ensure the highest quality of review and so that journal activity can be monitored and any trends – both positive and negative – can be identified and policy decisions made or appropriate rectifying measures put into place. All policy decisions, especially those that may have far-reaching and long-term consequences for the journal, should be based on reality and not on perception. Regular monitoring of reports may also

Table 2.1 Manuscript submission and resubmission 1991–2005 for Journal A.

Year	No. submitted	% increase over previous year	Mean submissions per month	% resubmissions
1991	169	–	14	–
1992	317	88	26	–
1993	374	18	31	–
1994	438	17	37	8.5
1995	438	0	37	7.5
1996	559	28	47	7.7
1997	632	13	53	5.1
1998	724	15	60	16
1999	703	–3	59	19
2000	738	5	62	18
2001	700	–5	58	20
2002	851	22	71	20
2003	877	3	73	22
2004	909	4	76	23
2005	1080	19	90	25

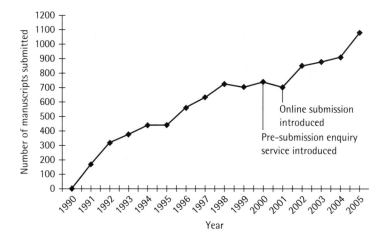

Figure 2.2 Annual manuscript submission 1990–2005 for Journal A.

Table 2.2 Geographical breakdown* of manuscript submission 1992–2005 (by country of submitting author) for Journal A.

Year	Australia and New Zealand	Europe	The East and Asia	North America	Other	Total
1992	9 (3%)	190 (60%)	19 (6%)	90 (28%)	9 (3%)	317
1993	11 (3%)	231 (62%)	29 (8%)	98 (26%)	5 (1%)	374
1994	15 (3%)	241 (55%)	46 (11%)	130 (30%)	6 (1%)	438
1995	19 (4%)	264 (60%)	34 (8%)	114 (26%)	7 (2%)	438
1996	11 (2%)	306 (55%)	65 (12%)	163 (29%)	14 (2%)	559
1997	26 (4%)	356 (56%)	69 (11%)	158 (25%)	23 (4%)	632
1998	31 (4%)	388 (54%)	108 (15%)	183 (25%)	14 (2%)	724
1999	21 (3%)	381 (54%)	110 (16%)	163 (23%)	28 (4%)	703
2000	15 (2%)	399 (54%)	130 (18%)	157 (21%)	37 (5%)	738
2001	23 (3%)	352 (50%)	124 (18%)	183 (26%)	18 (3%)	700
2002	23 (3%)	385 (45%)	180 (21%)	222 (26%)	41 (5%)	851
2003	14 (2%)	387 (44%)	222 (25%)	228 (26%)	26 (3%)	877
2004	22 (2%)	388 (43%)	195 (22%)	267 (29%)	37 (4%)	909
2005	31 (3%)	435 (40%)	273 (25%)	291 (27%)	50 (5%)	1080

*   Further breakdown within these geographical areas is also monitored.

Table 2.3 Mean time (days) from submission to decision – stage and period breakdown 2003–2005 from a report generated by an online submission and review system for Journal B.

Period	From author submission to submission centre	From submission centre to handling editor	From handling editor to two reviewers agreed	From two reviewers agreed to two reviews received	Time for handling editor to return recommendation	From handling editor recommendation to decision to author	Total time submission to first decision
2003							
Jan–March	2.4	0.5	11.6	20.7	10.0	3.9	43.5
April–June	1.8	0.9	13.9	20.4	11.2	3.8	44.6
July–Sept	1.4	0.9	9.9	18.0	9.2	2.6	37.0
Oct–Dec	1.5	0.9	11.5	19.0	8.2	3.0	36.2
Average	1.7	0.8	11.7	19.5	9.6	3.3	40.3
2004							
Jan–March	1.5	0.7	9.6	19.6	8.8	3.3	35.8
April–June	1.5	1.0	10.8	18.0	8.1	4.1	38.8
July–Sept	1.5	0.9	11.3	16.8	9.2	3.7	36.9
Oct–Dec	1.4	0.8	10.5	18.1	7.4	4.2	36.0
Average	1.5	0.9	10.6	18.0	8.4	3.8	36.9
2005							
Jan–March	1.5	1.2	9.1	17.9	6.8	5.0	37.1
April–June	1.6	1.0	9.6	17.3	6.6	2.9	33.9
July–Sept	1.6	1.0	12.3	17.4	7.3	4.8	37.7
Oct–Dec	3.3	0.9	11.4	18.1	6.4	3.3	37.2
Average	1.9	1.0	10.6	17.7	6.8	4.0	36.4

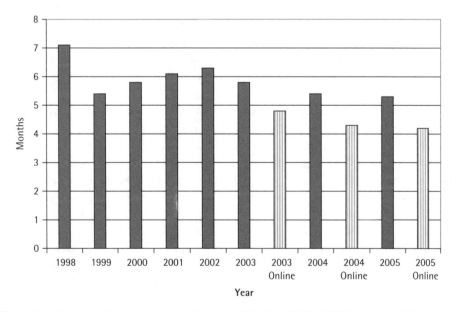

Figure 2.3 Average time from submission to publication 1998–2005 for Journal B.

enable a 'quick fix' to be put in place to resolve a problem rather than more drastic and permanent policy measures being introduced. Regular information dispersal to editors will inform decision making and help ensure it is consistent across the journal.

Systems available

Records can be kept in a variety of ways. Some small, specialist journals with a low number of annual submissions, and not anticipating much growth, may find that a paper system with some computer backup using ordinary office software for generating correspondence will be sufficient to cover their needs. Most editorial offices, however, now use computer systems to store information and manage the peer-review process. For small journals, a spreadsheet program can be used to record submissions, record transactions (such as when a manuscript is sent for review, to whom, when the report is due, when it has been returned and when the decision is sent to the author), and to monitor progress and check for outstanding reviews and decisions. Table 2.4 lists information that needs to be kept on manuscripts. Reviewer data can be kept in a similar way, and the sort of information that needs to be recorded is covered in Chapter 4 (see page 45).

Many journals, however, prefer to use systems that combine manuscript and reviewer information and manuscript tracking within a single searchable database. This makes things very much easier. Such systems can be developed individually, but unless you have the relevant expertise and/or good specialist and ongoing information technology support, this can be a tricky, and perhaps foolhardy, approach to take. Future editors and editorial office staff are also going to have to use the system, particularly if the editorial office is one that moves with changeover of editor-in-chief, and it is very likely that the new people may not have the expertise or support needed to deal with what has probably become a very individual system. It is far better to adopt a system that will withstand time and transfer. You also want to be sure you have a system that will cope with journal growth and technological advances in publishing, and not become unwieldy or grind to a halt once submissions reach a certain number.

Commercial stand-alone tracking systems are available, but the trend nowadays is to move to electronic or web-based submission and review systems. As more journals adopt these, authors are getting used to the ease and cost and time savings associated with them and there is increasing pressure from authors for journals to introduce such systems. A number of systems are commercially available. The market leaders have many hundreds of journals signed up with them and some now have more than one million users. There will, of course, be cost considerations, but if your journal is being published by a commercial or larger society publisher they may well have negotiated special discounted rates for their journals. They will also have the expertise in-house and with their journals to provide guidance and experience-based information on all aspects of online systems. They will generally provide help during the adoption and set-up period, and then support once the system is in use

Table 2.4 Information that needs to be kept on manuscripts.

Manuscript details
Required
Reference number
Title
Authors
Corresponding author (CA)
Address and contact details of CA
Type of article
Whether for a special or themed issue
Whether it is a revision
Whether it is a resubmission (with previous manuscript number)
Handling editor
Whether copyright assignment or licence to publish received if required
Whether all permissions and required correspondence received
Charges due
Any requests for waivers of charges
Any non-compliance with journal policy
Relevant notes

Optional (but recommended)
Length (word count or pages), possibly broken down into sections
Number of figures and tables
Number of colour figures
Whether supplementary material provided
Whether cover image submitted
Suggested and/or excluded reviewers from author

Transaction information
Required
Date manuscript received
Date manuscript assigned or sent to editor
Date reviewer list received from editor (if applicable to process operated)
Details of all individuals approached to review, with dates and outcome
Names of reviewers found
Dates manuscript assigned or sent to reviewers
Dates reviews due
Dates reviewers reminded
Dates reviews returned
Date reviews to editor
Date decision to CA
Decision
Date revision received (with above details if sent for review)
Date decision on revised manuscript to CA
Relevant notes
Current status
Date accepted manuscript to production and any relevant notes
Publication details for accepted manuscript

and on subsequent upgrades as the service providers develop the system. So, the best advice is to contact them or any publisher your journal may be moving to or thinking of moving to, to ask for advice and to find out what kind of service and support they will be able to provide. All the main service providers will be happy to demonstrate their systems to you, so contact them to arrange a session. Some have special arrangements for journals with relatively few submissions, so if you're a small journal or organization, don't feel you shouldn't approach the suppliers because they won't be interested or it will be too expensive. Moving to online submission and review is covered in Chapter 6, which describes how to go about choosing a system and then how to implement it. It also gives details on what to expect after introducing an online system.

With online web-based systems, you have a dedicated homepage for your journal, or group of journals if you are managing a stable of journals and want them linked in this way or to share a database. Editorial staff, editors, authors and reviewers are all able to access the site. Their access is user-name and password controlled and editorial staff can set up the system, or arrange with the supplier, to allow individuals access only to appropriate areas and manuscripts. Great care needs to be taken with online systems to ensure that all users are working on the correct manuscripts. If reviewers fail to do so, they will end up submitting the wrong report for a manuscript (see Chapter 4, page 66). If editors and editorial office staff fail to do so, they will be entering information incorrectly and communicating with the wrong people. This can have serious consequences (such as release of confidential information) as well as result in confusion. A simple tip is to introduce the '3N' Rule, whereby all editors and editorial staff always check three things before working on a manuscript or communicating with anyone via the online system: Number, Notes screen, and Name. This will ensure they're on the right manuscript screen, they've taken on board any relevant information before doing anything, and that they're communicating with the right person.

Whatever system is used to store information on manuscripts and reviewers and to manage transactions, the basic principles of what constitutes good practice in peer review remain the same. Also, whatever system is used, it is wise (although I accept some people may disagree with this) to have some paper record, however minimal. With electronic systems, paper records serve both as something to which to resort in the event of system or network failure, and as something to use in certain steps or checks because many people find it much easier to do these on a hard copy than on screen, particularly if they are dealing with a large number of manuscripts in a full-time capacity.

What kind of paper system is kept will depend very much on whether or not paper is the only medium in use and on personal preference and working style. In all cases, it is a good idea to create a hard copy folder for each manuscript. It's wise to invest in good-quality, hard-wearing folders that are large enough and won't fall apart prematurely, as some manuscript folders can become very bulky and may need to be consulted frequently. Don't buy large stocks until you're satisfied that the quality is right. Using heavy-duty rubber bands can be a great aid – this ensures everything is kept securely in the folders and they don't slip into one another, causing panic about

Box 2.4 Principles to observe when dealing with hard copy folders

- Never have more than one folder open on your desk. If you need to get another one out, for example because there's a telephone call from an author about his or her manuscript, put the other one onto another surface or onto the floor. It is all too easy to get sheets of paper mixed up and put into the wrong folders.
- Label all items with the manuscript reference number if it isn't already there.
- Put everything back in the folder when you've finished with it, in an order that everyone in the office follows, for example most recent material always being added to the front.
- Always return folders to the correct storage area after handling. Take care to move them on to the appropriate section if a new step has occurred.

'lost' manuscripts. The folder for each manuscript should be used to store everything relevant to that manuscript, and it itself is best stored in a filing cabinet. Folders can be colour-coded to denote type of article, handling editor, or whatever a journal considers to be most important or useful. At a minimum, the reference number should be on the front. A sheet with basic manuscript details such as title, authors, editor can be added to the front, and can even contain boxes that can be ticked to denote stage reached in the review process. The folders can be filed in numerical order so that any manuscript can be readily located. In this case, separate areas should be used for manuscripts that are active, ones that have been accepted, and those that have been rejected. The active area is the most important one for the daily work of the editorial office and so should be located where everyone who is going to need to use it can do so easily and without unduly disturbing other people. The accepted and rejected manuscripts will need to be accessed much less frequently, and not at all in some cases, and so can be kept in a less central area. There are some basic principles that should be observed when dealing with hard copy folders, as given in Box 2.4.

To introduce an element of manuscript management, the active manuscripts can be filed according to stage reached in the submission and review process, and ordered numerically within those sub-sections. For example, possible categories are:

- submitted, awaiting final check or additional material
- checked and complete, awaiting transfer to editor
- with editor, awaiting reviewer list or immediate decision
- reviewers being found
- with reviewers
- reviews all in, awaiting decision from editor
- authors revising

- revised manuscript in, awaiting decision from editor
- manuscript accepted, awaiting final materials.

This system makes it easy to run quick physical checks on manuscripts at various stages and to chase up authors, editors and reviewers who are late at any stage. With experience, staff will develop an 'eye' for how busy or slack different areas of activity are, and whether there are any bottlenecks that need attention.

Additional tricks can be used to help locate standard documents. Different-colour sheets can be used for things such as manuscript details, reviewer lists, or authors' suggested and non-preferred reviewers. These coloured sheets can be invaluable, especially in those folders that have grown very large and unwieldy, as they are very easy to spot. Care needs to be taken when lifting such large or joined multiple folders in and out of filing cabinets – they can be real wrist strainers (as well as occasionally inducing dread at needing to be dealt with again).

The paper record system needs to be evaluated periodically to make sure it still fulfils a need and is not excessive for the journal's requirements. There is no point duplicating everything that is stored digitally. It can be a wrench for some staff to abandon even parts of the paper record as it provides a comfort cushion on which they have grown dependent. Pruning back needs to go hand in hand with reassurance that various tasks can be done and information obtained without having to resort to hard copy information.

Paper folders will need to be cleared out regularly after manuscripts have completed review and been either accepted or rejected. How frequently this needs to be done will depend on how much storage space is available. Folders for accepted manuscripts don't need to be kept for more than a few months after publication. There is a difference of opinion on whether rejected manuscripts should be kept. Some editors feel they should be destroyed (although, with online systems, manuscripts remain available for reference and for generation of reports and statistics so this is not a matter of choice in this case). My personal preference is to keep rejected manuscripts, as information from their review will be relevant to the review of resubmissions and how the review process is handled (see Chapter 5, page 107). If hard copy folders are kept for rejected manuscripts, space considerations will usually dictate when they need to be discarded. If a resubmission comes in for a manuscript, the folder for the previous submission(s) should be retrieved from the rejected-manuscript storage area and attached to the folder for the new submission. It is important to make sure that all relevant important information from the earlier submission(s) is taken into account when considering the resubmission, so the folder for the original submission should be gone through carefully (along with the online record if an online system is used).

Care must be taken when disposing of files. They contain confidential and personal information and so their disposal must meet local data protection and handling requirements. If finances allow, a shredder is invaluable. If they don't, any sheets with confidential or personal information should be torn up or destroyed in some other way; they should not be put whole into waste bins or bags.

3 Manuscript submission and initial checks on completeness and suitability

All editors and editorial office staff hope to receive manuscript submissions that are suitable for their journals, are properly and well presented, with all the relevant information and enclosures supplied, and that have been submitted correctly. This makes everyone's life easier and helps ensure a more efficient and speedier review process. But to achieve this, authors need to know what a journal requires. Each journal's information or guidance for authors should therefore give very clear details on the journal's scope and policy, the types of articles it publishes, how authors should prepare their manuscripts, and how and where they should be submitted. If this information isn't readily, or clearly, available to authors, it's unrealistic and unduly optimistic to expect to receive manuscripts that are suitable, complete and consistent in format and presentation. Unsuitable or inadequately prepared or incorrectly submitted manuscripts can take up valuable editorial time, frustrating everyone involved, so it makes sense to do everything possible to help ensure manuscript suitability and complete and proper submission. Authors also have a responsibility to ensure that they submit manuscripts that are within the scope of journals, that they abide by those journals' policies, and that they follow their presentation and submission requirements.

An existing journal will already have instructions of some sort for authors. They may be good, or they may be bad. However adequate they are, they should be revisited periodically to keep them up to date with journal policy and technological developments. If a journal is experiencing problems with the suitability or presentation quality of manuscripts being submitted, the instructions need to be looked at carefully with a critical eye. Perhaps they aren't clear enough. Or maybe they're too lengthy and cumbersome, so that authors lose interest before reading them all the way through or can't be bothered to try to absorb all the minutiae. Experience shows that carefully-thought-out guidelines do lead to correctly submitted and well-presented manuscripts that are within the scope of the journal and can enter the review process without delay. A new journal needs to develop its own guidelines from scratch. If the journal is with an external publisher, contacts there should be able to advise and point to examples that work well for other journals. It's also worth checking out the information given to authors in a number of different journals. Editors will undoubtedly have experienced both good and bad guidelines when submitting their own manuscripts to journals, so their recommendations can be very useful.

Submission guidance to authors

Some of the things authors need guidance on will be journal specific, but many will be common to most journals. The guidance needed falls into three categories: journal scope and editorial policy, manuscript presentation, and manuscript submission.

Journal scope and editorial policy

It is vital that journals lay out their scope and publication aims clearly to deter, or at least minimize, unsuitable submissions. Such submissions will not only cause unnecessary work for the editorial office and editor, they may also result in annoyance or frustration for the authors, especially if it isn't clear to them that their manuscript isn't suitable, either because it's a type of article the journal doesn't consider or because its scope falls outside the journal's remit. How can the submission of unsuitable manuscripts be minimized? There are a number of ways.

Formulate policy
There should be regular editorial discussion amongst the editors to formulate policy on what sorts of manuscripts are not suitable. This should include actual subject areas and topics and also the minimum standards that have to be met for a manuscript to be considered by the journal. In fast-moving fields these may change quite frequently, as things that were once considered cutting edge and innovative become commonplace and standard technology.

Produce clear guidelines
The results of editorial discussions should be translated into clear and concise policy guidelines and be readily available and very visible to authors contemplating submission. The policies should also be very clear to all the editors of a journal so that they can act consistently. It's helpful to have a summary of editorial policy that they can refer to and this is also a valuable resource for new editors to help them become familiar with a journal and its requirements. The journal's peer-review policy should be made clear, particularly if open peer review is operated (see Chapter 4, page 42). It is also important that it is made very clear what the journal's policy is on preliminary reporting and prior publication, and what will exclude a manuscript from consideration, for example prior publication in conference proceedings or posting on a public server. Confusion by authors on this may inadvertently lead to unacceptable duplicate publication (see Chapter 9, page 178). Many journals provide no or very little information. But some bodies, for example the American Physiological Society, provide quite detailed guidance on what, for them, constitutes 'prior publication' (see www.the-aps.org/publications/journals/apsethic.htm).

Have a pre-submission enquiry service
A very good and efficient way of screening out clearly unsuitable manuscripts is to offer a 'pre-submission enquiry service'. If authors are unsure whether their

manuscripts are suitable for a journal, they can submit an abstract along with a cover letter describing any special considerations or concerns. With email and an efficient and organized system in place, these enquiries can be dealt with easily and quickly. They can be readily circulated to the most appropriate editor(s) and most authors should be able to expect a response within a few days. Since all the manuscripts that receive negative responses on pre-submission enquiry might otherwise have been submitted as full submissions, having this kind of service in place can result in very considerable time and work savings. It also leaves editors and reviewers free to concentrate on manuscripts that are genuinely suitable for consideration.

Experience shows that the majority of such papers will probably not be suitable. Some will be, though, and submission can be encouraged to whatever degree the journal wants. Others will fall into a grey, 'possibly, but not sure' category, either because it is impossible to assess suitability without seeing the whole manuscript, or because a manuscript falls at the borderline of what is acceptable for that journal. In the latter case, it pays to be honest with authors and to tell them exactly this and the reason, or reasons, why. In particular, if a journal is about to change, or it is in the process of changing, its editorial policy or raising the thresholds for acceptance, authors need to know this, and also what the new criteria will be and, if appropriate, what additional work they would have to include to meet the new standards. If editors are unsure about any manuscript but would be willing to give it the benefit of the doubt, authors can be told that if their manuscript were to be submitted it would be sent out for full review but that the prospects for a positive outcome are not good. In both these instances, the decision then lies with the authors, and it is up to them if they want to invest the time and effort needed to get the submission to the standard required and/or the time that would be taken up with a full review process. Sometimes they may have no possibility of carrying out additional work – the project may have been terminated or the relevant people may have moved away to new jobs. A lot will obviously also depend on whether they suspect they are close to being beaten to publication or whether they have the luxury of time. In the former case, they may decide to trade off journal ranking against speed and go for a lower-impact journal where they can be reasonably sure of getting their work accepted. Editors are often knowledgeable about what other journals require, in terms of both scope and standard – they may well have acted as reviewers for those journals, been on their editorial boards, or just be keeping up to date with what those journals publish – and it can be very helpful to authors for them to suggest alternative journals that would be suitable for submission of specific manuscripts. Authors are frequently very grateful to get this sort of advice. But editors should refrain from referring to 'lower-quality journals' as this may understandably be offensive both to the authors and to the editor of the new journal. A more appropriate description should be used, for example 'more specialized journal'.

It is good practice to develop a reference number system for pre-submission enquiries and to ask authors to cite this number when/if they do make a formal submission. This will allow information relating to the pre-submission enquiry to be linked to the full submission. As some submissions may take some time to come in, for example if authors are told that their manuscript would be suitable if more experiments were included, the reference number should, to avoid confusion, incorporate the year. A very simple example could be: PS06-008, standing for pre-submission enquiry number 8 in 2006. The potential editor's initials could also be added, as authors can be given guidance on editor choice for their submissions if the journal asks authors to select an editor. They can also be advised of any specific enclosures or supporting material that would need to be provided with a formal sub-mission and this will help ensure a proper and complete submission.

Manuscript presentation

Authors need to be given information on various aspects of the presentation of their manuscripts. The International Committee of Medical Journal Editors (ICMJE) has established guidelines for the formatting of manuscripts submitted to biomed-ical journals and they are known as the 'Uniform Requirements for Manuscripts Submitted to Biomedical Journals: Writing and Editing for Biomedical Publication' (www.icmje.org). These originate from an informal meeting of a small group of edi-tors of general medical journals in Vancouver in 1978. The group became known as the Vancouver Group (which evolved into the ICMJE) and the guidelines are some-times referred to as the Vancouver Guidelines. These were first published in 1979 and have been updated regularly ever since. The Uniform Requirements now cover more than just manuscript preparation, as they have been expanded to include ethical considerations and various publishing and editorial issues. They are well worth a look at by editors of all scientific journals, and by the authors submitting to them, irre-spective of whether or not they are biomedically related.

Manuscript format
If there are any length restrictions, on either text or displayed material, these should be clearly stated and an indication given of whether these are absolute or subject to editorial discretion. Do word and/or character counts need to be provided, and if they do, does a breakdown for each manuscript section need to be given? Guidance on language style, nomenclature and use of abbreviations should be given, along with any reference sources that can be consulted, either generally available ones or documents that might be available specifically for the journal, perhaps from its or its publisher's website (if it has an external publisher). It can be helpful to give exact details of what information the top page of the manuscript should contain: for example, full title of the paper, the short running title (and whether there is a char-acter number restriction for either), the full names of all the authors, the name(s) and address(es) of the institution(s) at which the work was carried out, the present addresses of any authors if they are now different, the name, address and contact details of the author to whom correspondence should be sent (called 'corresponding author' throughout this book), keywords, accession numbers, word and/or character

counts. If a journal operates a peer-review system where the authors' identity is kept from the reviewers (masked or blinded review – see Chapter 4, page 41), then authors will need to be advised on how to prepare their manuscripts to eliminate identifiers. The title page can contain all the details described above – as it is useful to have them together in one place – but this needs to be additional to the first page of the manuscript the reviewers will see, which should not contain the authors' names or affiliations. The manuscript should not contain any running headers or footers that might identify the authors. Authors should be advised to refer to references to their own work in the third person, not the first, i.e. to avoid phrases such as 'as we have previously shown (Jones *et al.*, 1995)', and use instead the format 'as Jones *et al.* (1995) have shown'. Care should be taken that authors do not inadvertently identify themselves in the Acknowledgements section, for example by referring to grants awarded to named people (even initials should be avoided, as in many cases they will be enough to identify the authors). Those names must, however, be added before publication, because details of funding sources and to whom the grants have been awarded should always be provided. For paper submissions, the top page of the manuscript can be removed prior to sending to the reviewers; for electronic submissions, authors will need to send separate files for the cover page and a version of the whole manuscript suitable for the reviewers. Care must be taken to eliminate any identifiers to the creators of the electronic files if the authors have not done so.

Beware!

If author identity is being kept secret from reviewers, check that all identifiers have been removed from electronic files. Make sure the authors don't give away their identities in the Acknowledgements section of their papers.

The order in which sections should appear should be specified, and whether new pages or separate files should be used for any of these. All these format issues will affect the ease, efficiency and thoroughness of review; a well-prepared manuscript puts editors and reviewers in the right frame of mind and allows them to get on with the real job of reviewing.

Figures and artwork

Full and clear instructions should be provided on figure and artwork preparation, both hard copy and electronic, as required by the journal. Help given to authors here will save much time later and ensure that figures of high enough quality for review are submitted. If reviewers receive inadequate figures they may not be able to assess a manuscript properly and might put it aside, perhaps never to return to it, or they may submit an incomplete review. Some publishers provide detailed instructions for electronic artwork preparation (for example, Blackwell Publishing; www. blackwellpublishing.com/authors/digill.asp). Some online submission systems also run checks on figures being submitted and alert authors if the quality is inadequate.

Reference details required

Journal reference style should be indicated, both for method of citation within the text and for format within the reference listing, along with examples. It should be made clear if full titles of papers cited are required. These are important in the review process as they can alert editors and reviewers to similar or related papers that have been published and that might need to be considered in the review of the submitted manuscript. Failure to include full titles in submitted manuscripts leads to unnecessary delays while authors are requested to prepare and submit them prior to review or to reviewers not being able to carry out a proper assessment because of lack of possibly relevant information.

Enclosures required

Clear details should be given of all the enclosures needed. If any actual signatures are required then it's very important that authors know this so they can start getting these together some time before they are ready to submit. Scientists travel a lot, go to conferences, go on field trips and take sabbaticals, so if all signatures are required it may take some time for these to be collected. Examples of other enclosures are: copies of related manuscripts in press or submitted elsewhere (see Chapter 4, page 65), and how many copies are needed; copyright assignment or licence to publish agreements; permissions to reproduce items from published material (note that these can take considerable time to obtain); permissions to cite unpublished work or personal communications (and whether actual consent letters from the people concerned are required – this is a good idea and ensures that this information is not misrepresented, either in error or intentionally, or included without the knowledge and approval of the individuals providing the information; see Chapter 9, page 181); agreements that the authors will respect the policy requirements of the journal; all ethical committee and licensing authority approvals and personal consents; CONSORT (see Appendix II, page 217, and Appendix III, page 270) or other checklists and flowcharts; any identification or registration numbers and registry details that are required, for example for clinical trials, public database depositions, materials depositions; letters indicating any deviations from journal policy or restrictions on availability of materials, or giving explanations for any requests, for instance why certain people have been requested to be excluded from review; details of all funding sources; disclosures of any potential conflicts of interest.

Journals dealing with clinical trials should provide clear and up-to-date guidance on their requirements, for example registration in a named public registry prior to submission and provision of an International Standard Randomised Controlled Trial Number (ISRCTN; http://isrctn.org, see Appendix III, page 272). Standards for clinical trial registration are in the process of change, so readers to whom this is relevant are advised to consult the website of the World Health Organization (WHO; www.who.int/ictrp/about/details/en/index.html), which is spearheading the initiative and attempting to introduce greater regulation and transparency into clinical trials with the formation of the International Clinical Trials Registry Platform (ICTRP). The ICMJE is also working towards the establishment of a comprehensive publicly available database of clinical trials and provides very useful information about this on its website, including answers to frequently asked questions about clinical trial registration (www.icmje.org/faq.pdf).

Number of copies required

If submission is on paper, how many copies of each manuscript (and the various enclosures) need to be provided? Also, do all sets need to contain photographic images or is one 'original' copy sufficient, with good-quality laser printer copies for the other sets?

Supplementary material

Supplementary material can be of two kinds:

1   Material that is to be used to aid assessment by the editor, or by both the editor and the reviewers, and to be reviewed along with the manuscript under consideration, but which is not intended for publication. It may be supporting data or statements, or additional images that strengthen the claims being made in the paper. The reviewers or editor may after review advise authors that it, or some of it, needs to be included in the paper itself.

2   Material that is submitted for consideration for publication but that is to be published only in the online version of the journal, not in the paper version. It is an integral part of the paper and so should be subjected to peer review. Such material can be anything from raw data sets and additional figures or tables to movies and data analysis software. The data sets can be very large (and are rapidly becoming very much more so in certain disciplines such as molecular biology) and so can only realistically be published online. It is important to note that it is crucial that the material be kept available with the published paper for all time, not only so readers can evaluate the work presented in the paper, but also so future readers can revisit the data as science advances and assess them in the light of future discoveries or ask different questions of them.

Clear guidelines need to be given on how this supplementary material should be submitted or accessed, especially as it can be very complex or require specialist techniques, and special arrangements may need to be made for its review (see Chapter 4, page 63). Authors need to be helped to ensure that the supplementary material that is intended for publication is in a format that can be handled both by the editorial office during review and by the production process.

Manuscript submission

Mode of submission

Authors need to know whether to submit on paper, by email attachment or via an online submission site and whether there is a choice or one mode is mandatory. If any method is explicitly forbidden, for example by fax, this should also be stated if this is a problem.

Place of submission
Authors need to know where they should submit – to a central office or direct to a regional or subject handling editor.

Editor choice
If submission is via a central office, authors need to know if they can select a preferred named editor whom they would like to handle their manuscript. If they can, is this optional or obligatory?

Assistance with submission
Authors may have queries about their actual submissions or experience problems during a submission, particularly with online submissions. It's therefore helpful to give the person or people, and their contact details, with whom they should communicate to get help or more information.

Checking and logging of submitted manuscripts

Whether manuscripts are submitted on paper or electronically, the same basic checks need to be made. If at all possible, it's a good idea to print out online submissions. They're much easier to read and check than on screen, especially if comparisons or checks are being made against other manuscripts, and there will always be a hard copy available to work from if the local network or online system is down. However, this is something that is down to personal preference, and some people may feel this is not necessary to the way they like to work.

It's very useful to have a standard checklist sheet that editorial staff can use when logging in manuscripts. This helps ensure that all essential things are checked and it also makes it easier for those members of staff who do not regularly deal with this area to step in when needed and do an efficient and complete job. An example of such a checklist sheet is given in Appendix II. All journals will want to check some of the things on this, but some will not consider everything is necessary. Others will also have other specific requirements that need to be met before a manuscript can be put into the review process and these should be added to their checklist. The checklist needs to be updated as technology moves forward and affects submission procedures and standards, and when editorial policy changes and authors are required to submit additional items or to submit things in a certain format(s).

Each journal needs to decide how much time and effort can be spent on the initial checking of submitted manuscripts and to identify the critical elements. It depends very much on how much assistance is available. However, it is good practice to try to resolve all manuscript deficiencies before moving a manuscript on in the review process. Time spent doing this at the beginning can save a great deal of time later on and prevent editors and reviewers from getting frustrated because they don't have all the correct items or these are of insufficient quality for them to be able to carry out a proper review.

All new manuscripts need to have their receipt acknowledged with the authors. This serves two main purposes: firstly, it lets the authors know that their manuscript has arrived safely and gives them a date for establishing priority for their research results; secondly, it provides the opportunity to give the authors a reference number, which they should be asked always to use when corresponding with the journal about that manuscript. This not only speeds up dealing with queries, it also helps prevent mixing up of manuscripts – which can be a real danger if a group of authors has more than one manuscript submitted or there are manuscripts with similar titles or author names submitted or in the review process. There must be a systematic manuscript numbering system in operation. This is automatic with online and most commercial stand-alone tracking systems – the numbers are allocated sequentially as manuscripts are submitted or entered. But if a journal has developed its own electronic or paper-based system for manuscript tracking, it is crucial that it also develops an adequate reference numbering system – one that will stand the test of time, so it needs to feature the year, the manuscript number and, if more than one journal is being handled or is likely to be dealt with by that office in the future, a journal identification. For example, the third manuscript submitted in 2006 to Journal A could be identified as A-2006-0003. It's a good idea to put in the zeros in the final part to maintain correct numerical order in listings.

The acknowledgement of receipt is also a very useful point at which to request missing or inadequate materials, and appropriate sentences should be added to the acknowledgement communication if the system in use allows. If it doesn't because an automatic acknowledgement is generated by the system, a separate email can be sent after the manuscript has been checked. It's a good idea to build up a set of standard sentences that will cover all eventualities and which all staff can use (Box 3.1 gives some common examples). Authors will usually respond quickly, especially if they are told that their manuscript cannot enter the review process until the missing or improved materials, clarifications, and so on have been received. In reality, some of these manuscripts can enter the early stages of the review process, in the expectation that any requested items will shortly be received. Minor errors in format won't matter, but missing pages, figures, tables or copies of related manuscripts submitted or under consideration elsewhere are serious omissions that will hold up review and so need to be provided before review. Authors will very soon learn what a journal expects and what they can and cannot get away with. If a journal builds up a good reputation for insisting on high-quality presentation for its submissions, authors will be wary of trying to submit inferior material and reviewers will also be more prepared to review because they know they are going to receive manuscripts that are complete and well presented.

Making sure that manuscripts are complete and well presented before they enter the review process should be the general rule in the offices of journals where

> **Box 3.1 Examples of standard sentences to ask authors to supply missing or inadequate items**
>
> 'We have noticed on checking your manuscript prior to transferring it to the editor that Figure X is missing. Please would you send this to us as soon as possible so that processing of your manuscript is not delayed. You can email it to us at. . . .'
>
> 'We have had problems opening the file for your supplemental Figure X. Please would you send us a new file as soon as possible so that processing and review of your manuscript is not delayed. You can email it to us at. . . .'
>
> 'It is journal policy that all related manuscripts in press or submitted elsewhere must be sent with submissions. We have noticed that Smith *et al.* from your group is in press at another journal but has not been included. Please would you send a copy of this to us as soon as possible so that the review of your manuscript is not delayed. You can email it to us at. . . .'
>
> 'The journal has manuscript length restrictions and requires that all authors give details of the word counts of their manuscripts, including a breakdown for each section. This information is missing from your submission. Please would you send this to us as soon as possible so that processing of your manuscript is not delayed. You can email it to us at. . . .'

manuscript receipt is dealt with by editorial office staff who do not have decision-making powers. However, in small journals or where all new submissions are received or immediately seen by an editor, a distinction can be made for those manuscripts that are clearly unsuitable for the journal and are going to be returned without review. There is no point, for example, getting an author to submit improved-quality figures if the manuscript isn't going any further. A note of caution: there are some very efficient editors who deal with new submissions very rapidly and their decisions for rejection without review may be made very soon (a few hours or same day) after actual manuscript submission. If authors receive rejection responses that quickly, they can sometimes feel that no one has really read their manuscript, even though someone has and has done so thoroughly, and so they may be a little aggrieved. Waiting a day to let authors know the outcome can be good psychology! The very short delay will make no difference to the author in terms of time lost, and it may well avoid doubt and inaccurate assumptions by the author and possibly appeals for reconsideration.

Transfer to editor

When a manuscript has passed the initial checks and is complete it needs to be passed on to someone who has authority to make editorial decisions. For some journals, particularly those that don't have large numbers of submissions, all manuscripts may be seen by the editor-in-chief, who will then decide either to handle the manuscript him- or herself or to pass it on to another editor (if there are other editors). If the editor-in-chief is located in the same place as the editorial office, it's just a matter of passing a copy to them if hard copies are still being used. If the editor-in-chief is distant, and a paper-based system is still in use, a copy will need to be mailed to them, which may take some time, as very few journals can afford to use courier services as the norm. With online systems, transfer is quick and easy, and the editor-in-chief can have looked at a manuscript, made the decision on who should handle it, and assigned the manuscript to that person with access made available in a very short space of time. Choice of editor will depend on subject matter and current workload (both of which are very easy and quick to check with online systems), and on whether any editor may already be handling a similar manuscript, in which case they should generally also handle the new one (unless there are any conflicts or contraindications as described in the following paragraph).

New manuscripts may be assigned to editors based on author choice. This is a common system and can work well, especially for a journal that has a large number of submissions. The advantages are that all manuscripts get assessed quickly and, theoretically, by the most appropriate editor. Also, as editors get to see all the manuscripts for which authors have selected them, they won't miss getting those manuscripts they may have been encouraging authors to submit to the journal. Journals with author selection of editor should make sure that editors feel they can request transfer to another editor if a manuscript falls outside their area of expertise or if they are too busy to take it on. The disadvantages of author choice for editor are that some editors may get far too many manuscripts and may start to feel overwhelmed, even though they know they can request transfer to another editor. An initial screen by the editor-in-chief or managing editor to weed out those that are definitely not suitable for the editor, or need to be excluded for some reason, for example if the authors come from the same institute or an editor has requested they not handle papers from certain individuals or groups, will reduce that problem. If a journal does ask authors to select an editor, it's a good idea to put a 'let-out' clause into the submission guidelines so that authors (and editors) don't complain that a different editor from the one selected is handling their manuscript; for example:

> The journal reserves the right to assign the manuscript to another editor depending on, for example, workload, subject area, or extended absence.

Beware!

Authors intentionally choosing the wrong editor with the expectation of an easy review.

A note of caution: some authors may intentionally choose the 'wrong' editor, one who doesn't seem appropriate, exactly because of their lack of knowledge about the subject of their manuscript! They may feel that this will give them a positive advantage in review because the editor may not send the manuscript to the most appropriate reviewers or recognize flaws in the manuscript. This strategy can, however, backfire, as lack of relevant expertise or knowledge on the part of the editor may mean that they miss something which would be recognized by a specialist as being very significant but which to a non-expert appears not to be very important. Editors should be encouraged to return to the editorial office for reassignment manuscripts for which authors may have chosen them inappropriately. A great advantage of online systems is that multiple changes of editors can be made in a very short time, and each potential editor can view a manuscript and all the relevant information they need to make a decision about whether they will handle it or whether another editor would be more appropriate. A significant amount of information, some high level and very important to the review process, can be gained from this editorial process and should be noted accordingly and be available for reference while manuscripts are being considered by the journal. If an online system does not create a permanent record of which editors have commented on the manuscript, this information should be added to the notes screen.

Some authors will request that a certain editor be excluded from handling their manuscript. This can be a tricky situation and is certainly one that needs to be handled diplomatically. There may be very valid reasons why the excluded editor should handle the manuscript; they may, for example, be the only editor with the relevant expertise or they may be handling related manuscripts. In that case it is good practice to let the author know and give them the opportunity to withdraw the manuscript. In other cases, the authors may have a very legitimate reason to request exclusion and the journal may choose to go along with the request, particularly if there are other editors who can perfectly well take on the manuscript. If a journal finds that it has an editor whom authors are repeatedly requesting be excluded, it may be a time for a parting of the ways. It is not good for a journal's reputation to have an editor whom his or her scientific peers are avoiding, and it's also not fair on the other editors, who may begin to resent having to struggle with manuscripts outside of their direct areas of expertise when the journal actually has an editor who is a specialist in that area. Such situations need to be investigated to determine the reasons behind the requests for exclusion. If it is because that editor's acceptance standard is too stringent compared with the other editors, that can be addressed and resolved. But if it is because of concerns over conflicts of interest or possible misconduct (see Chapter 9, page 183), punitive action will need to be taken if these prove to be founded (see Chapter 9, page 192).

Initial assessment of suitability and rejection without external review

Each newly submitted manuscript should go through an initial check to make sure that its content falls within the scope of the journal, that it follows editorial policy guidelines, and that it does not contain an unacceptable level of overlap with manuscripts by the authors that are either in press or under consideration elsewhere. If this doesn't occur, and all manuscripts are automatically sent out for review, there is the danger that you will end up with angry reviewers and/or authors. Reviewers may be annoyed because they have spent time, and perhaps resources, on manuscripts that are clearly not suitable for the journal and feel that this assessment should have been the responsibility of the editors. There will always be some manuscripts that slip through, and reviewers will understand this. But if it happens regularly, reviewers may start to feel that they are doing the editors' work for them and end up disgruntled and unhappy with the way the journal is operating. You will also be tying up your reviewers with manuscripts that are not suitable when they could be reviewing ones that are, which is wasteful all round. Authors may well be annoyed that the review process has been unnecessarily prolonged, and feel their manuscript should have been returned to them following an editorial assessment if the reason given for rejection is unsuitability for the journal. A full review process of perhaps a couple of months may mean that they are scooped by competitors publishing similar work elsewhere.

The percentage of manuscripts being returned without external review varies from journal to journal. It is very high (around 50%) for the top, high-impact general science journals, which receive many thousands of submissions each year and cannot possibly review them all. Many papers reporting sound results are returned simply because they are not novel or significant enough to be of interest to a wide audience. For journals with lower submission numbers, or those in specialist areas, the percentage will depend primarily on how many manuscripts fall outside the scope of the journal or are clearly of poor quality. Some may be found on an initial scan to have too much overlap with work already published or submitted elsewhere. All journals will also, from time to time, receive some strange or even weird manuscripts. Giveaway signs may be: having a single author for a research paper (which in many scientific disciplines is unusual nowadays, although does still occur), submission from a home address, or the presence of a bizarre or grandiose title.

The mechanism for rejecting manuscripts without external review also varies from journal to journal. For smaller journals, the editor-in-chief may have sole responsibility, or may consult informally with others. Some journals give authority to the individual handling editors to make these decisions on their own or after consultation with other editors. Other journals have regular meetings where candidates for rejection without review are discussed. If there is not consensus on rejection, manuscripts may be put into the review process and sent for external review.

The decision letters to authors whose manuscripts are rejected without external review should be polite and state clearly exactly why their manuscripts are being rejected, with specific reasons (see Appendix II for examples). Not only is this courteous to the authors, it also serves to educate them so that, over time, scientists in a journal's community will recognize the type and scope of paper suitable for that journal. Reasons for rejection of a manuscript might include any number of the following: it has insufficient originality, it is too specialized, it is of limited interest to the journal's audience, it contains serious scientific or methodological flaws, the topic is not covered in enough depth, it is interesting but too preliminary and/or too speculative. If the work has the potential to be interesting to and accepted by the journal, the opportunity can be taken to provide advice on what would be required for it to be put out for full review, and to add whatever level of encouragement is appropriate.

Although manuscripts can be rejected without involving external reviewers, they cannot be accepted for publication without external review. All papers published in reputable scientific journals must have been peer reviewed, and peer review must involve assessment by external reviewers (see Golden Rule 2). This should hold however 'hot' and worthy of publication an editor feels a manuscript to be. Experience shows that editors can be mistaken about such manuscripts, with problems coming to light when external reviewers are brought in.

Manuscripts with language problems

There will be some manuscripts submitted whose content is suitable for review but in which the standard of language is poor. These need to be looked at carefully and an accurate assessment made of the extent of the problem. If the language has deficiencies but the text can be relatively easily understood and the results interpreted without great difficulty, then it will probably be acceptable to send the manuscript out to reviewers. It is good practice to mention to reviewers that the journal is aware that there is a language problem and that this will be corrected at revision or during copyediting if the manuscript is accepted. An apology might be appropriate, with a note saying that if the reviewer finds that it becomes too difficult or burdensome to assess the manuscript accurately they should let the editorial office know and seek advice on how to proceed.

Beware!
Sending manuscripts to reviewers in which the standard of language is very poor is unfair and may frustrate or anger them.

If the standard of language in a manuscript is very poor, then it should not be sent out to reviewers. If it were sent, reviewers would have to struggle with it and it would be highly likely that they would not be able to understand and interpret the results and conclusions, and so not be able to provide an accurate assessment of the work. They may also experience frustration and get annoyed, feeling that they have wasted their time and that the journal is asking too much of them. This can all be detrimental to the work under review and prejudice the reviewers against it. In such cases, an editorial assessment of suitability for the journal should be obtained. If the journal would be interested in seeing the manuscript resubmitted, the authors should be contacted to explain that the manuscript is being returned to them for language improvement and that this is being done in their best interests, as it is felt that it cannot receive a fair review in its current state. The journal would, however, be willing to put out for review a version in which the language problem had been addressed. If it is decided that the manuscript is not suitable for the journal and stands little chance of being made so, the authors should be notified of rejection without external review with the recommendation that the manuscript be submitted to another journal. They should also be advised to improve the language before submission anywhere else as, in its present state, the manuscript's chances of being well received by any reputable journal are greatly reduced (see Appendix II for sample letters for these two situations).

For manuscripts submitted from countries or laboratories where there are likely to be native speakers of the language, you can suggest to the authors that they ask one of their native-language speaking colleagues to go over and correct the final version of their manuscript. In countries where this would be difficult, and perhaps financial resources are limited so using a language correction service is not feasible either, journals should be prepared to be more accommodating and use discretion on the threshold for acceptance for language quality. However, for any manuscript that has a co-author whose native language is that in which the manuscript is written, there is no excuse for authors to submit work that is badly or inadequately written – that author should take responsibility for language correction, and this should be made clear to the corresponding author.

4 The full review process

After weeding out unsuitable manuscripts, i.e. those that haven't for one reason or another passed the initial assessment on suitability (see Chapter 3, page 36), and obtaining all missing items or better-quality versions of inadequate ones for those manuscripts that do warrant further review, you are left with manuscripts that are appropriate and are in good enough shape to be sent to reviewers. To constitute proper peer review, assessment *must* involve external reviewers, as stipulated by Golden Rule 2. This external review will involve:

- identifying and selecting appropriate reviewers
- finding reviewers
- getting the manuscript and associated material to the reviewers
- helping the reviewers with any problems they may encounter during review
- monitoring review progress and ensuring reviews are returned in a timely fashion
- receiving and checking of returned reviews
- getting the reviews and ancillary information ready for assessment and decision by the editor handling that manuscript.

The external review process can be of different types and it is important that the system a journal operates is clear to both authors and reviewers in its editorial policy guidance, as some individuals may choose not to submit to or review for a journal if they are not happy with the system in operation. Peer review can be 'closed' or 'open'.

Closed peer review. This is the traditional system, and still the one used by the majority of journals. In closed peer review, there is anonymity involved. If the authors do not know who the reviewers are but the reviewers know who the authors are, this is termed 'single blinding'. In 'double blinding', the identities of the authors are kept from the reviewers as well as the reviewers' identities being kept from the authors, so the authors and reviewers are unaware of each other's identity. The reviewers also do not know who the other reviewers are. Some journals have introduced double blinding in an attempt to reduce possible bias, whereby reviewers might be influenced, either positively or negatively, by the status or reputations of the authors and their institutional affiliations, or just by knowing who they are. However, a real problem is that it is difficult to achieve successful blinding of author identity. Authors can be asked to make sure that their manuscripts do not contain any identifiers, and editorial staff can take care not to include any identifying information in the manuscript and documents they send to reviewers (see Chapter 3, page 29), but despite this, a number of studies have found that around half of reviewers can guess author identity.[1] They may get enough clues from the citations in the manuscript and/or from knowledge of the work going on in that field. They are, after all, experts in that area. The suggestion that blinding reviewers to author identity leads to better opinions and reviews isn't really borne out by trials.[1-4] As blinding involves a good deal of work on the part of both the authors and the editorial staff, journals need to consider carefully whether it is something they want to do, especially as some people argue that blinding removes information that can actually be useful to reviewers in making a thorough and complete review. The term 'masking' is sometimes used interchangeably with blinding. The latter (blinding) may, however, refer to keeping author identity secret and the former (masking), if it is used, to keeping reviewer identity secret (both from the author and from the other reviewers). But as the reverse can also be found in many publications, it is a rather confusing situation and the best advice is for a journal to choose which terms it will use, to define these, and then be consistent in their use.

Beware!

The terms 'blinding' and 'masking' are sometimes used to denote the same thing and sometimes different things; confusion abounds.

Open peer review. In open peer review, the authors know the identity of the reviewers. Some editors and journals, particularly in the biomedical field, have become concerned that a closed system of review is open to abuse and that it is wrong for people making important judgements on the work of others to do so in secret and so not be accountable for their opinions and recommendations. They argue that the system should be completely transparent. The following reasons have been put forward by Godlee for the introduction of open peer review[5]:

1   It is more ethical as there is greater accountability for both editors and reviewers, and less scope for bias or the misappropriation of ideas and/or data by reviewers or undue or deliberate delays in them returning their reviews.

2   There don't seem to be any adverse effects on the quality of the reviews or the usefulness of the reviewers' comments.

3   It is possible to put into practice.

4   Reviewers can be given public credit for their work. Their names can be published with the papers, and with the actual reviews if the journal has extended open review to include their publication. The opportunity is presented for editors to grade reviews and for the best to be highlighted, thus reflecting positively on those reviewers.

The *BMJ* (*British Medical Journal*) introduced open peer review in January 1999, largely for ethical reasons as controlled trials by the journal had shown that asking reviewers to consent to being identified to the authors had no important effect on review quality (it didn't lead to higher-quality opinions, but nor did it lead to poorer-quality ones), the recommendations on publication, or review time.[3] There was a significantly increased likelihood of reviewers declining to review, which may be a concern for some journals contemplating moving to open peer review. A common argument against open peer review is that it may lead to animosity, reprisals and damaged relationships. Junior researchers, for example, may be reluctant to make honest criticisms of work submitted by senior figures in their field, fearing that this might have negative repercussions on future grant proposals, applications for jobs, promotion prospects, and so on. The willingness of researchers to accept open peer review will likely vary across disciplines. Editors know their journal communities and so are best placed to gauge how receptive they will be to the introduction of open review in their specific areas. You could therefore consider polling your reviewers to see what their reaction would be, and whether significant numbers would decline to review in future under such a system. If you do decide to introduce open review, it would be wise to monitor review quality and recommendations to see whether these are affected in any way, and also whether it is more difficult to find reviewers, especially for manuscripts that are more problematical.

Identifying and selecting appropriate reviewers

The first stage in external review is to identify and select suitable reviewers. Reviewer selection is one of the most important aspects of the peer-review process, arguably the most critical. The most appropriate reviewers need to be identified for each manuscript. An inadequate manuscript sent to reviewers who are not knowledgeable in the area, or who are notoriously 'soft', and handled by an editor who is also a non-specialist may end up being accepted for publication. This is unfair. Such manuscripts cannot really be said to have been 'peer reviewed'. Journals have a responsibility to ensure that all manuscripts submitted to them are reviewed appropriately and rigorously; no manuscript should ever be given an easy ride. Reviewers have a duty to assess manuscripts objectively, and always to remember that they are reviewing the work, not the authors (see Golden Rule 6).

Golden Rule 6

Reviewers must assess manuscripts objectively and review the work, not the authors.

To help ensure rigorous review, a record or audit trail needs to be kept of where the various reviewer suggestions have come from for each manuscript, and all relevant reviewer-related information must be kept with each manuscript. One way to do this is to build up a 'Reviewer information' section for each manuscript; this can prove invaluable and greatly enhances the quality of review. For paper submissions, it can be a summary sheet attached to the file. Most online systems have a notes screen for each manuscript and the reviewer information can be held there, for example at the bottom of the other notes, in a dedicated block that can be easily referred to when necessary (see Box 4.1). Alert points should be highlighted in some way, for example within asterisks or in capitals. The information is then available to editors and editorial office staff and can help inform decision making on reviewer selection and the fate of the manuscript.

The potential reviewer database of a journal needs to be large enough to ensure that manuscript assessment and publication decisions are not in the hands of just the same few people. What is considered a 'large' database will vary from journal to journal, depending on how general or specialist they are, and between subject areas. They can range from many tens of thousands of individuals for large multidisciplinary journals to just a few hundred for specialist or niche journals.

Reviewer information

All journals should keep a database of their reviewers and this should be maintained and kept up to date. The database may be shared by a group of related journals. Established journals will most probably already have a database, either paper or electronic, and its quality and usefulness will depend on the rigour and professionalism of the previous editors and editorial office staff. It cannot, though, be assumed that all journals do have such a database. If its full extent is that it exists only in someone's head (which, though hard to believe, is not unknown!), then you need to start from scratch. It is alarming that 18% of the respondents in a survey on peer review carried out by the Association of Learned and Professional Society Publishers (ALPSP) and the European Association of Science Editors (EASE) in 2000 said they did not keep any records on reviewers.[6] If you're inheriting a journal and it does have a database, you need to look at it carefully and make sure it contains all the sorts of information it should. Also, check if you can whether it has been subject to regular housekeeping and updating, and what the system for this was. Think carefully about what information you want to ask of and keep on reviewers. If only paper records exist, it is advisable to migrate to an electronic system and you will need to work out how much of the existing information you actually want to input. It may be easier to input information only as you start to contact and use reviewers, using the paper system as a guide for that, and you can also be sure then that all the information in your database is up to date. You need to keep information on some or all of the following for reviewers.

Personal data

1   Full name and title, including preferred mode of address if there is one. Note especially if this is different to the given name, for example if a nickname, or Western forename for non-Western reviewers, is preferred. The family name also needs to be established as this is not always clear with authors of different nationalities. For some disciplines, it may be important to record professional qualifications.

2   Current contact details, including postal address, telephone and fax numbers, and email address. If the person has more than one address, you need to record when each is to be used and whether this is a recurrent situation, for example for annual work at a field station, or just a one-off instance, for example for a term of sabbatical leave.

3   Whether the person has, or has had, any official role with the journal, for example as an editor or advisory board member, and the dates these positions started or were held.

4   Periods when the person does not want to receive manuscripts for review. This should cover both regular times such as annual examination periods, and single or infrequent instances, for example sabbaticals, parental leave, or absence due to personal or family reasons.

5   Areas of interest and expertise, with, if possible, some indication of the level of knowledge and familiarity with these various areas (see this chapter, page 48). A note should be made if the person has any unique or unusual combination of knowledge, skills or expertise. It's important that the information kept on people's interests and expertise is that which will best help the journal identify specific reviewers.

6   Whether there are any problematical issues, either in areas of review or if there are reasons not to use the reviewer in any situation(s) or for certain manuscripts. But take care! It should always be borne in mind that reviewers may ask to see data that are kept on them. All journals need to ensure they comply with the data protection and storage laws in their countries to ensure data are accurate, relevant and secure, and all legal requirements are fulfilled. Any information that might be construed as damaging or unfair should not be kept on such a database.

7   Special requests made by the person, for example not to be asked to review manuscripts from certain individuals or groups.

8   Whether there are contraindications to the involvement of the person in the review of manuscripts from any individuals, for example those related by blood or marriage, who have been recent supervisors or supervisees, or with whom there are serious conflicts of any sort (see Chapter 8, page 166).

Reviewing activity

1   Current workload of manuscripts for review.

2   Number of manuscripts reviewed in the past, dates they were done, and the manuscript reference numbers.

Box 4.1 Typical 'Reviewer information' section of the notes screen of a manuscript on an online system

****

REVIEWER INFORMATION

Non-preferred/excluded reviewers: D. Jones (bad personal relations with author) and A. Smith (recent collaborator).

Editor wants 3 reviewers, one def to be Guy Willing (offer longer time).

ASK reviewers to comment on soundness of methodology. Make related MS 751-2005 available to all refs.

Ref 1 (AG) suggestion from Editor PM.

Ref AG declared potential conflict but Editor happy for him to review.

ON HOLD REVIEWER - Sandra Weiss (can review any time - OK just to send MS).

9/11/05 Poppy Soon: "Sorry. My baby is due in 3 weeks, and I'm declining new assignments. As alternative reviewers, I suggest:

    May B. Available   mbavailable@......

    Will Help   will.help@......."

9/11/05 Ben E. Away "Sorry I am out of town until 15 December."

11/11/05 Ref Willing: "I accept to review this ms but can only start when I return from a trip to the US, on Nov 20 and will need 3 weeks."

12/11/05 - Ref 3 (GW) needs longer - Send Date of Reminder 1 changed to 11 Dec.

24/11/05 Ref AG (1): "Message received - will send you my report tomorrow."

29/11/05 Ref Help (2): "will send detailed review to you tomorrow - manuscript looks very nice. Should have only very minor suggestions for the authors... Sorry to be a bit slow but with the Thanksgiving Holiday here in the US last week everything is running behind. - Will"

****

3   It's useful to keep a note of the number of times an individual has been asked to review but declined. Some online systems will record this automatically. This information is helpful when assessing level of activity, possibly with a view to deciding on people to invite to become an editor or to join a journal's special reviewing panel. Level of activity may be low for two reasons. Some people may regularly decline invitations to review, and not just for those manuscripts with which they have potential conflicts, which isn't a healthy indicator of commitment to the journal. However, there may be reviewers whose level of activity is not high because they aren't often asked to review, perhaps because they are in a niche area of expertise. They may offer specialist knowledge that is very valuable to the journal, and when they do review they may always do so willingly and do an excellent job. These are positive indicators that need to be considered alongside actual numbers of manuscripts reviewed.

Reviewing standards

1   *Timeliness of review.* This is an objective measure and it is easy to input information on how good a reviewer is at returning reviews in the agreed time. All reviewers have periods when circumstances prevent them returning their reviews on time, but every journal should have a good idea of how fast or slow its reviewers generally are. The editorial staff can easily enter the 'timeliness' assessment to help alleviate editors' workloads and this can be done as reviews are received.

2   *Quality of review.* This is a subjective assessment, best made by the editor handling the manuscript, who will know the area and be able to grade a review accurately on usefulness, thoroughness, and so on. Length of review is not an indicator of quality, as a long report may be just a list of relatively trivial comments or language corrections. Conversely, a short, concise report from a real expert highlighting, for example, major flaws, will be of great help to the editor in making a decision on a manuscript. Online systems frequently have the facility for recording review quality, and editors should be encouraged to enter this information when dealing with manuscripts as this will help build up a comprehensive reviewer database that will be of assistance not only to them, but to other editors and staff trying to identify reviewers for manuscripts.

For information on reviewers to be helpful it needs to be kept up to date. This will happen to some degree on an ongoing basis, with people voluntarily notifying the journal of changes of address or other details, either by direct communication or by logging onto a journal's online site and entering the changes themselves. However, many reviewers won't think to notify journals of such changes and so journals need periodically to contact them to check. Occasional dedicated mailings – by regular mail or email – to all reviewers on the journal's database need to be made. The most efficient way to do this is either to send reviewers their details as they appear in the journal's records and ask for changes to be indicated and the record returned, or, for online systems, to ask the reviewers to visit their online account and update their details there themselves. For journals still using paper copies, address checks can be made when reviewers are invited to review by including the current address held and asking for confirmation of accuracy when they reply. This will avoid manuscripts being sent to wrong addresses. Care should be taken not to overload reviewers with too many requests for updates, for example by doing it every time they review, as there is the danger that they will get fed up and start to ignore the requests. It is something that might be done annually, but how practicable this is will depend both on the size of the reviewer database and on the time available. If data entry and updating are done by the journal, it is good practice that the person entering new details onto the database or updating information initials the record and notes the date data were entered or amended.

Beware!

Don't overload reviewers with too many requests for updates to their personal and professional details.

Data on reviewers who are inactive for one reason or another need to be kept on the database, both to avoid adding their details afresh and inadvertently sending new manuscripts to them, and to have a permanent record of their reviewing activity. The reason for inactivity needs to be recorded. There will be reviewers who are inactive because they are no longer asked to review due to problems with fairness, reliability or review quality. Others will be inactive because they no longer, for a variety of possible reasons, wish to review for the journal – they may get very annoyed if they are contacted, and relations with the journal, if they were poor before, may deteriorate further. There will also be individuals who have died; this needs to be recorded to avoid causing any embarrassment or distress by sending out manuscripts or invitations to review to reviewers who are deceased.

How to go about identifying and selecting reviewers

How are reviewers identified and selected? Journals differ in how they do this. Ideally, reviewer selection should involve more than just a mechanical semi-automated process of matching up manuscripts with reviewers based on titles and keywords. Some journals may, however, do this, and if they do, it is crucial they develop a comprehensive and accurate keyword database for their reviewers' areas of interest and expertise. This needs input from people knowledgeable about the subject areas covered so that terms that mean the same thing can be consolidated and appropriate lists developed. Crucially, the terms need to be those that are most useful to the journal. Reviewers can be given lists of keywords to choose from (the Royal Society of Chemistry, for example, provides extensive lists of keywords, broken down into categories, for reviewers to select those keywords most appropriate to them; www.rsc.org/Publishing/ReSourCe/index.cfm, and see Appendix II, page 243), referred to a subject headings database (such as, for example, the Medical Subject Headings of the US National Library of Medicine, MeSH®; www.nlm.nih.gov/mesh/), or they can be invited to submit their own. With an automated keyword-matching system of reviewer selection, there is always the danger that manuscripts will not go to the best-qualified reviewers – authors can be very bad at giving their papers good and appropriate titles and keywords, and some important aspects of the work reported may be missed out completely or the correct emphasis not recognized (the article by Hartley and Kostoff[7] is a good exposition on the usefulness or otherwise of keywords in scientific journals). Reviewer identification and selection is a stage at which human specialist knowledge is invaluable. Consequently, many journals, especially those striving to achieve high quality and excellence, involve editors here who are specialists in certain areas. Journals rely on such specialists to provide the depth and insight that are needed to identify appropriate reviewers, and to bring new names to the reviewer database. This is one of the reasons they will have been appointed as editors. They will be aware of the people in their field, and will likely know or have met or have experience of a large number of them. They will know what work is being done and by whom, how the field is developing, and which areas are problematical. They will also be aware of the rivalries that exist between various groups and of any animosities, and whether they are bad enough to prevent manuscripts from certain individuals or groups ever getting a fair review from certain other groups. In the ALPSP/EASE survey of peer review in 2000, for many of the respondents the reviewers were predominantly selected by the editors (78%) or by members of the editor's team (43%).[6] Other means were used by relatively few (14% nominated by the author, 8% by computer, and 7% by members of the publisher's team). Twenty-nine per cent used more than one method.

The editors may be the people who contact potential reviewers and arrange review, but for many journals, especially the larger ones and those with centralized offices, this is done by the editorial staff. Editors are required just to send lists of suggested reviewers to the office staff, who then carry out all the finding of reviewers, sending manuscripts out to them and chasing up reviews. The following steps need to be taken whoever is responsible for finding reviewers for manuscripts.

A list of potential reviewers needs to be compiled
The editor should read the manuscript and accompanying material, and identify potential reviewers in the following ways:

- from the journal's database
- from his or her own contacts or suggestions from those individuals
- from his or her knowledge of scientists in the field
- from the bibliography given in the manuscript
- from literature and database searches
- from suggestions put forward by the authors.

Authors may volunteer names of reviewers or submit them in response to a journal's invitation to do so. If author suggestions are invited, guidelines should be given on who it is not permissible to suggest; for example, it is common to exclude anyone who has been a co-author, mentor or advisee in the past few years. It is not usually necessary to state that relatives or colleagues in the same groups or institutions should be excluded, as these are quite clearly conflicts of interest (see Chapter 8, page 166). However, some journals may, even so, want to do so, or find they need to. Different journals will have different criteria both on who should not be suggested and on how long ago the association should have existed to become no longer relevant. Academic associations cannot generally exclude people indefinitely, as there may not be enough people with the right expertise left to review! Care should be taken with an author's suggestions. They may be totally valid and appropriate. However, a number of journals will have experienced receiving suggestions of reviewers who are close friends of the authors or who are not even experts in the area, and it is up to the editor to try to spot these inadequacies. Some suggestions may result from a group of authors wanting to increase the visibility and importance of their area in the published literature and forming 'cabals' whose members review one another's work favourably. Again, specialist editors should be able to spot these and bring in a reviewer(s) from outside that area if appropriate. There is also some evidence that, although author- and editor-suggested reviewers do not differ in the quality of their reviews, reviewers suggested by authors tend to make more favourable recommendations on publication and are more likely to recommend acceptance.[8,9] If an author-suggested reviewer is used, editors may therefore want to balance this with another/others selected by the journal.

Beware!

Be alert to groups of authors in a small specialist field reviewing each other's work favourably to increase the visibility of that area.

Authors mention people in the Acknowledgements section of their papers whom they would like to thank or acknowledge for various reasons, such as being involved in discussions about the work, for reading the manuscript and providing suggestions for improvement, for data collection or providing materials, and so on. Most will be genuine, and it is totally appropriate and right that all such help should be acknowledged. However, astonishingly, it is not unknown for authors to include the names of people who have never seen the manuscript or had anything to do with the work! One can only guess that this might be an attempt to add credibility to the work or might even be a ploy to try to ensure that those people aren't approached to review the manuscript as they have been involved with the work and might have a potential conflict. Either scenario is unethical.

To make the final reviewer selection for the list of potential reviewers, editors need to take into account any contraindications that exist, such as intense rivalries between groups, or groups working on exactly the same topic with possibly work either under consideration for publication or close to submission. If journals allow authors to put forward 'non-preferred' reviewers or those whom they would like excluded, then it is up to them whether to respect this in each case or to suggest that those reviewers are involved in the review. It is good practice to ask authors to state the reasons why they have put forward non-preferred or excluded reviewers (and these should be added to the notes or notes screen for that manuscript; see this chapter, page 44, and Box 4.1); these will help editors decide what to do and if any extra measures need to be put in place if such a reviewer is used. Reasons may range from potential bias due to close association through co-publication or joint projects or grants, to suspicions of severe conflicts of interest and consequent fears about an unfair review. The latter may be genuine fears based on previous experience, but they may also be totally unfounded, and frequently unwarranted, suspicions. However, there is evidence that manuscripts where authors' requested exclusions are honoured have higher acceptance rates.[10] If a non-preferred reviewer is used – there may be good reasons why that person's opinion is really important – editors may want to use an extra reviewer just in case that person's assessment does prove to be suspect or unduly harsh, or even (perhaps intentionally) unacceptably delayed. A note of caution about allowing authors to list non-preferred reviewers: it is wise to stipulate how many they can submit, as some authors may end up listing so many that virtually everyone with the relevant expertise is excluded! Two or three names are reasonable in most cases.

Beware!

If authors are allowed to list non-preferred reviewers, make sure the number allowed is specified to avoid the requested exclusion of a large number of experts.

It can save time if editors can provide lists of five to six reviewers at the outset, so that whoever then goes on to deal with finding reviewers for manuscripts does not need to keep getting back to them for extra names and so introduce delays, and possibly frustration. The chances of achieving a high quality of review are greatly enhanced if editors can also provide the following information with their lists of suggested reviewers:

- whether there is an order of priority
- whether a certain person should definitely review the manuscript, preferably with the reason for this (that reviewer may need to be approached as a special request if they are already reviewing or currently not reviewing because of other commitments or work overload)
- whether specialists from different subject areas or with different expertise are required; and if they are they should be listed in groups according to these areas

- whether the reviewers need to be given specific instructions, for example asked to concentrate on certain aspects or to comment on specific points
- whether the reviewers need to be provided with additional material, for example given access to another manuscript under review or in press, that might have a bearing on review.

Appropriate notes on the above need to be added to the manuscript notes or notes screen (see Box 4.1).

Checks that need to be made on reviewer lists

Whoever is searching for reviewers will need to carry out certain checks on reviewer lists received from editors. They need to check:

- the suggested individuals are not authors on that manuscript (quite a few authors will be able to tell stories about times they have been asked by journals to review their own manuscripts)
- the suggested persons are not in the same groups or at the same institutions as the authors
- there are no contraindications, for example close relationships, previous requests from any one of the individuals not to be asked to review manuscripts from that group, or they've submitted bad or biased reviews in the past for these authors
- each individual's current and past workloads, and whether maximum manuscript review number in a certain period has been reached (either by reviewer request or according to journal policy)
- whether the individuals are currently being asked, or have just agreed, to review other manuscripts from the journal (or its sister journals if they share a database). If reviewer searching is not centralized, it is very important that all the people who are involved in doing this can make central checks. This is very easy and straightforward to do with online systems; journals with other systems need to put appropriate procedures in place. Reviewers understandably do not take kindly to being asked by different individuals to review a number of different manuscripts for the same journal (or even for a group of related journals which share a database) at the same time.

If non-preferred reviewers have been included in the list of suggested reviewers, editorial staff may sometimes want to check back, diplomatically, with the handling editor to make sure he or she is aware of this and whether they want to make any special arrangements for review, such as using an extra reviewer. Staff will know their editors and be aware which are very thorough and observant and which may sometimes miss things and so tend to need, and usually appreciate, gentle reminders.

Number of reviewers to be used

Peer review requires that external reviewers are used (Golden Rule 2), but how many reviewers are consulted varies. Many journals elect in most cases to go to two outside reviewers. An ALPSP survey in 2005 found that 75% of the journals used two reviewers, 17% three reviewers, 2% a single reviewer, and 6% used more than three reviewers.[11] These results agree closely with those found in the ALPSP/EASE peer-review survey of 2000,[6] where 73% of the respondents typically used two reviewers, 18% used three, 6% used a single reviewer, and 3% used more than three reviewers. Sometimes additional reviewers are required if a specialist in a particular area is needed, for example a statistician, or there is a suspected conflict of interest with potential review bias with one of the reviewers. In some journals, one external reviewer will be used together with someone from the journal's editorial team. Choosing the number of reviewers is a fine balance between ensuring a thorough review and not overloading reviewers. It must never be forgotten that reviewers are a very valuable and precious resource (see Chapter 7) and should not be abused or used inappropriately.

Finding reviewers

Once potential reviewers have been identified for a manuscript, actual reviewers have to be found.

To invite or not?

Some journals send manuscripts to reviewers without first asking them if they would be willing to review. In the ALPSP/EASE peer-review survey of 2000, only about half of the respondents (54%) contacted reviewers in advance to ask if they would be available.[6] Those journals that don't first contact reviewers may feel that it would be too much of a burden for editorial staff or editors to invite reviewers and that they don't have the time needed to do this. However, there are several drawbacks to not first contacting potential reviewers:

- It is rather discourteous, and an imposition on that reviewer unless that person has given a blanket instruction that the journal can send manuscripts without first checking.

- A reviewer may be away or have moved, and so a journal may be waiting fruitlessly for a review to be returned. If paper copies are still being used, that copy of the manuscript may be lost to the journal. This could cause problems as there may not then be enough copies for other reviewers, an important consideration in cases where reviewers need to receive original high-quality images to assess the work properly.

- A reviewer may return a manuscript unreviewed for a number of reasons, varying from being too busy to having a serious potential conflict of interest. Valuable time will have been lost and the review process delayed. Some reviewers may not return a manuscript until a journal sends the return postage costs, again introducing delays.

- If a reviewer hasn't agreed to review, they are not under any obligation to return a timely review, or to return one at all, and the manuscript may be assigned to the bottom of their 'to-do' pile.

- The individual may not be the right person to review – they may not be familiar with the work or they may have moved away from that research area since last reviewing for the journal.

There are great advantages to contacting potential reviewers and getting their agreement to review ahead of sending the manuscript:

- It is courteous to check that a reviewer can, and is happy, to do the review.

- The reviewer can check that the manuscript falls within his or her area of expertise and let the journal know immediately if it does not.

- If the reviewer agrees to do the review, and to doing it within the stated timeframe, then the journal has every right to remind and chase him or her to return the review once that date has come and gone.

- The reviewer may provide useful feedback before, or instead of, doing the review. He or she may have reviewed the manuscript before, or know of problems associated with the work and recommend that certain people should review it. It is up to editors and editorial staff to evaluate such information and decide whether it is to be acted upon, but it can considerably enhance the quality and depth of review.

- If a reviewer is too busy to do the review themselves, he or she may suggest names of other suitable reviewers.

It is therefore recommended that potential reviewers should, if possible, be contacted and their agreement to review obtained before sending them a manuscript. Journals may well find that concerns they have about extra time being needed to do this do not prove founded because the review process becomes much more efficient.

Inviting reviewers

If invitations to review are sent they should be well thought out and contain information that will help a reviewer to decide, firstly, whether they have the necessary expertise to assess the manuscript, and, secondly, whether they can do it in the timeframe expected. The invitations might routinely include: manuscript reference number, title and (unless the journal operates author-blinding; see this chapter, page 41) author listing; the type of article (review, research paper, and so on) and if it is to be part of a special issue; the time for review (which may be dependent on paper type as many journals will allow longer for review or opinion articles than for research papers, but require a rapid review for 'fast-track' papers); and whether any special arrangements are being made for the review, for example if extra time is being allowed – this may help get a positive response, especially from regular reviewers who will be expecting the normal return time. The following is an example of wording included when finding reviewers for a special issue:

> We would normally ask our reviewers to return their reviews to us within 10 to 14 days. However, we realize that reviewing may take a little longer than this to provide a full assessment of manuscripts for the special issue, and so we have extended this to 21 days.

Consideration should be given to what other information is relevant or should be included to persuade a certain reviewer to accept the invitation, for example that the editor would particularly value their opinion (and reason why if available and appropriate to pass on), or that the manuscript is related to another they are reviewing or have reviewed, or it is a resubmission of a manuscript they reviewed that was not accepted on first submission. If the reviewer is already reviewing a manuscript, or more than one for the journal or its sister journals, including a note of apology and explanation as to why they are being contacted again so soon will always be appreciated by reviewers, and they may be more likely to agree and not feel they are being taken advantage of.

Journals need to decide how many invitations to send out to reviewers at the same time – a lot will depend on the number of potential reviewers and the response times that are usual in the community covered by the journal. In fast-moving fields or those with populations who are very reliant on information technology and are often sitting at their computers, response times may as a matter of course be rapid. This also applies wherever they are in the world, as people nowadays tend to pick up emails and voice mail very regularly even when travelling. Inviting more people than needed is fine, but sending a manuscript to more than are needed in the expectation of using only the first two reviews returned is not good practice and should most definitely not be done. It is very discourteous, as the other reviewers may be investing a lot of time and effort in their reviews and, with good reason, be upset to find when they come to submit them they are no longer needed. Next time they are approached by the journal to review, they may on principle decline and good reviewers may be lost. Or they may accept an invitation to review but then decide that they won't after all submit a review, or take a long time, in the knowledge that if they leave it long enough their review won't be needed anyway. If only the first reviews received are used, there is also the danger that these might have been done very quickly and may consequently be superficial. If a journal bases its decisions on the first reviews returned from a larger pool of reviewers, what happens if any submitted after the decision has been finalized and sent to the authors bring to light flaws or present information that would have made the editor return a different decision or ask for different revisions if that information had been available earlier? It can get very messy and result in an inadequate review process.

How long a journal should wait before sending out more invitations to review while still waiting for replies from people already contacted will depend on a number of factors: for example, knowledge of a person's usual response time, how quickly a manuscript needs to be reviewed, the time of year (things often take longer during vacation times), how important it is that certain individuals review a particular manuscript. If the list of potential reviewers is reasonably long, then it is sensible to move to the next person on the list. But if there were very few names to start with, or the list has been virtually exhausted without success in recruiting reviewers, then individuals already contacted but who haven't responded need to be chased up. In these cases it can save time to send the original request again, or the information in it, with the new message as individuals may have lost, or never have received, the original message and accompanying information. Some may reply saying they thought they had already replied.

There are two situations where journals may experience much quicker and much slower responses (or no response). If potential reviewers themselves have a manuscript under review with the journal, they may be very keen to stay in favour and readily accept to review. They may then either review quickly or hold on to their review and may even attempt to initiate a bargaining situation (see this chapter, page 72). If a potential reviewer has just or very recently had a manuscript of their own rejected by the journal, they may not be inclined to review, or even to respond, especially if they feel that they themselves did not get a fair review or that the decision was wrong. They may even very honestly declare that they do not feel they could at the present time review a manuscript objectively and without bias. Reviewers are normal human beings and so are subject to all the usual human sensitivities and failings as well as the virtues. This should always be borne in mind and situations dealt with sensitively, and allowances made for people who occasionally act out of character because of difficult or disappointing circumstances.

Responses to invitations to review

People approached to review may respond in a variety of ways. All relevant information from the responses, and from any ensuing correspondence, should be added to the manuscript notes or notes screen.

Yes, they can review

1   Yes, they can review, and in the timeframe required (the response all journals hope for).

2   Yes, they can review, but will need longer. The journal needs to decide whether the extended return time is acceptable. Knowledge of the usual review time of such reviewers is important, as some reviewers never return their reviews on time even if they agree to a timeframe, but others who say they will need longer may be very reliable about keeping to any extended deadline.

3   Yes, they can review, but want to involve a co-reviewer. If this is someone in their own research group, for example a senior post-doctoral fellow, this can

work very well as it frequently results in a detailed critical review combined with the balanced overview that comes from the additional knowledge and insight that experience brings. Reviewers should, however, be made aware that they should not contact anyone else about reviewing a manuscript without the knowledge and permission of the journal. The usual checks need to be run first (see this chapter, page 52). If the request to involve a co-reviewer extends to an external colleague the editorial staff may want to check back with the handling editor that they are happy with this arrangement. Some won't be, for any number of reasons, and it is better to find this out right at the start of the review process than at the end when the reviews are in and an editor says they will not accept a review from that person and they want another one. With time, as staff get to know each new editor and their personal preferences, they will be able to establish procedures for dealing with their manuscripts and know when they need to refer back to an editor and when they don't.

4   Yes, they can review, but can cover only part of the work as they don't have the expertise to assess the whole manuscript; a reviewer expert in area X is also required. The editor may already be aware of this and have taken this into account in reviewer selection. If this is not apparent from the reviewer list, and there isn't anyone in the editorial office who is experienced or knowledgeable enough to provide guidance, the editor needs to be contacted with this information so that they can suggest new reviewers to approach if necessary. Feedback should also be given to the original reviewer as appropriate: that the editor would still like their opinion and that reviewers expert in the other area(s) will also be assessing the manuscript, or perhaps to thank the reviewer for bringing this to the journal's attention and that in the circumstances other people will be approached to review. Reviewers should never be afraid of letting a journal know about limitations to their ability to review certain manuscripts; editors will value this feedback because they can make appropriate adjustments to their reviewer choice to ensure that manuscripts are properly assessed. Sometimes, a general overview is required from a non-specialist who is perhaps a trusted scientist whose opinion an editor respects and values. If this information is known at the start (editors should be encouraged to provide this as a matter of course when they submit their reviewer lists; see this chapter, page 51), it is helpful to mention it to the reviewer when inviting them to review as it can save a lot of going back and forth between the various parties.

5   Yes, they can review, but they have reviewed it before, for another journal. Some editors may have very definite views on whether or not to use such reviewers. Similarly, some reviewers may feel strongly that they should be excluded from the new review to give the authors a second chance, with fresh reviewers. In some cases, a reviewer may be biased against the manuscript to start with because of the previous history of the manuscript – there may have been rebuttals and appeals, perhaps challenging the reviewer's competence – resulting in them not being able to assess it objectively the second time round. If reviewers

do not want to review a manuscript for a second time, or do not feel comfortable for whatever reason doing that, their wishes should be respected. In cases where a reviewer is happy to review a manuscript again, it's helpful to keep an open mind and assess each instance on a case-by-case basis. There are pluses and minuses to consider. Having the same reviewer assess the manuscript again means they can tell whether it has been genuinely improved since they last saw it. They can also assess whether any worrying, and perhaps unacceptable, changes have been made in an attempt to deal with the previous comments and criticisms, for example by leaving out, or even changing, problematical data that might question or change the interpretation of the results. The following quote from a reviewer's confidential report to the editor demonstrates this:

> I reviewed this manuscript for . . . a few months ago and it has not improved much since then. There were some data in the previous form of this manuscript that led me to seriously question the interpretation and significance. The authors chose to simply delete those data for this version of the manuscript. Hence, I have even less confidence in the work now than I had previously.

Reviewers may sometimes say that in view of the severity of their original criticisms and the timing of the submission to a new journal, it would have been impossible for the authors to have dealt with their criticisms adequately in that time. However, if the same reviewer is used again, it should be made clear to them that they should assess the manuscript submitted to the current journal; they should not assume it is unchanged from the earlier submission to another journal and so they should not base their review solely on their earlier assessment, and should certainly not submit the same report. This is not good practice. If a person approached to review sends their previous comments uninvited, editors may or may not choose to ignore them. If they decide to take on board the information it should be used only as ancillary information; it should not be used to prejudice an editor and prevent a manuscript getting a fair review, but rather to inform reviewer choice if appropriate.

6   Yes, they can review but have a potential conflict of interest. The decision on whether to proceed depends on what this conflict is, how the reviewer feels about it, and whether the journal considers it to be of relevance. If the editor feels he or she would still like to have that reviewer's opinion despite the potential conflict and the reviewer is happy to provide a review, then there is no problem. The potential conflict of interest is again ancillary information that can be used in the decision-making process (and so it should be noted in the notes or notes screen for that manuscript). The editor may, however, decide to involve an additional reviewer to counter the effect of any potential conflict. If a reviewer feels at all uncomfortable about reviewing a manuscript despite an editor being happy for them to do so, their feelings should be respected and they should not be pressured into reviewing the manuscript.

7   Yes, they can review, but the reply is received too late, after a sufficient number of reviewers have been found. Such people may prove to be very useful further

on down the line if problems are encountered – for example, if one of the reviewers does not return their review and another reviewer needs to be found quickly (see this chapter, page 71), or if divided opinions are received from the first reviewers and the editor would like to have another opinion (see Chapter 5, page 91). A brief message should be sent to the late-responding reviewer thanking them and explaining the situation, saying that the journal would like to keep them as an 'on-hold' reviewer in case their help is needed. A note on this needs to be added to the manuscript reviewer record (see this chapter, page 44, and Box 4.1).

No, they can't review

1   No, they can't review because they are too busy and don't have the time. They may indicate a date after which they could review. This should be noted, as some manuscripts are very difficult to find reviewers for and reviewers may still be being sought when that date is reached and so the manuscript may need to be sent to that reviewer.

2   No, they can't review because there is a problem of some kind. The problem may be personal and unrelated to the manuscript, in which case an appropriate note needs to be added to the individual's record if it means they should also not be contacted about any other manuscript at the present time, and for how long. If the problem is related to the manuscript, the editor should be made aware of it and whatever action is appropriate taken. The 'problem' could be trivial or it could be very serious and one that demands immediate action, for example if the reviewer has at the same time received the same manuscript or a very similar one for review from another journal. It is unethical to submit a manuscript to more than one journal at the same time. In cases where this principle has been contravened, editors should stop the review process immediately. They should seek an explanation from the authors and notify all parties about what has occurred. Contact with the editor-in-chief of the other journal will be needed if simultaneous submission to more than one journal (so-called 'dual or multiple submission'; see Chapter 9, page 179) is proven; in such cases, Golden Rule 3, which states that the submission of a manuscript and all the details associated with it must be kept confidential, needs to be broken. Serious problems and suspected or alleged misconduct must always be investigated – they should never be ignored (see Chapter 9, page 173). It is very important, however, that all accusations or suspicions are independently substantiated before passing them on to anyone else or any punitive action is taken (see Chapter 9, page 190).

3   No, they can't review because, as a matter of principle, they never review manuscripts from that group. Some reviewers, because of bad experiences, personal animosities, and so on, make it their policy never to review work from certain individuals or groups, and may request that they never be contacted to review for them. Such requests should be respected and noted on the reviewer's record for future reference.

4   No, they can't review because they themselves have a similar manuscript submitted elsewhere or in preparation. Potential reviewers should declare a conflict
of interest if asked to review a manuscript that is very similar to one they have
submitted elsewhere or have in preparation for submission. Occasionally, the
person may suggest to the journal that they also submit their article in preparation and if both manuscripts make it through the review process they be published back-to-back (i.e. consecutively in the same issue of the journal; see this
chapter, page 82, for how to deal with back-to-back manuscripts). Some editors
will be receptive to this, especially if the two manuscripts complement one
another and it will enable the journal to publish a much more complete picture.
Sometimes, people contacted in this situation will decline to review but the
awareness the request creates will lead to them submitting quickly to another
journal to try to beat the other authors to publication. Many journals have been
on the receiving end of such letters of submission; it won't be clear how the
authors have got to know of the related manuscript, but it can be guessed that
some have learned about it through a request to review. It would be highly
unethical for such a potential reviewer to request to see the competing study,
supposedly with a view to deciding whether or not to review, but if all they have
done is receive an unsolicited invitation they cannot really be criticized too
harshly for wanting to get their work published fairly soon. Editors and journal
staff should, however, be alert to reviewers who accept a manuscript for review,
receive it, and then return it saying they have a conflict of interest or aren't
really expert enough to provide a review. These may be genuine reasons, but
some may be cases of reviewers wanting a sneak preview of a manuscript for
personal reasons, which is totally unethical. This highlights how important it is
to keep accurate and complete records of who has been asked to review a
manuscript and what their response was (even, or rather especially, if they
declined to review; see this chapter, page 44, and Box 4.1).

Getting the manuscript and associated material to the reviewers

Once a reviewer has agreed to review a manuscript it should be sent to them,
or instructions given on how to access it, as soon as possible. If any delays are
anticipated, the reviewer should be notified so they can adjust their schedule if
need be, as they may have set aside a specific time slot for the review. Reviewers can
be given access to manuscripts in a number of ways, depending on the system used
by the journal: via a hard copy sent by regular mail, by email attachment, or by
downloading from a web-based online system. Whatever the medium, the general
principles and good practice are the same. In all cases, reviewers should receive
manuscripts that are correctly formatted, well presented and complete, with all the
materials they need to be able to carry out a proper and in-depth review included.

They will need clear instructions on review, information sheets, checklists and reviewing forms.

Information reviewers will require

Examples or sources of items referred to below are given in Appendix II.

Instructions

Instructions need to be given on how to access the manuscript if it is not being sent by regular mail, when the review needs to be submitted, and how that should be done. A contact name and/or telephone number and email address should be provided for reviewers to use if they have problems or need further information or assistance.

Information sheets

Information should be provided on journal policy, with guidance on the sorts of papers the journal is seeking to publish. If certain types of paper are no longer being accepted, these need to be specified. The journal's policy on specific requirements needs to be spelt out, especially if the reviewers are expected to look out for and comment on them. Information sheets should be reviewer-friendly and not so extensive or complex that reviewers will tend to ignore them completely.

Checklists and reviewing forms

Most journals have standard checklists and/or reviewing forms that they like their reviewers to fill in and submit as part of their review. These can help reviewers greatly by focusing their attention and acting as reminders about certain aspects that need comment. These checklists and forms are, except for the report intended for the authors, nearly always solely for editorial use and not transmitted to the author; reviewers should therefore be advised if they are not confidential, as most will assume they are. Unless all information is passed to the authors, reviewers should be reminded to keep separate information intended only for the editors and that which can be forwarded to the author. Failure to do this can cause problems, both because inappropriate comments may get transmitted to the authors and because arrangements need to be made for such comments to be moved to the correct place, thus causing extra work for editorial staff and possibly delays (see this chapter, page 75). Reviewers should also be reminded that they should not, either in their confidential comments for the editor or in the comments meant for the author, make derogatory, inflammatory or libellous personal statements. They should be instructed that their narrative reports written for the authors must correspond to what they have indicated in the confidential parts of their reviewing forms and checklists. The two sections of the review should be consistent or there may be problems when relaying editorial decisions and reports to authors. Some journals prefer that reviewers do not include any recommendation on acceptance or rejection in their reports for authors; in these cases reviewers need to be told about this as many will be used to including recommendations when reviewing for other journals. Some reviewers may object to reviewing forms and be willing to submit only a written report. Such reports from some reviewers can be of tremendous value and those reviewers should not be pushed into filling in forms if they really don't want to. However, if reviewers say they want to provide comments only for the editor, not for the authors, then they should be persuaded to do so as it is not fair to the authors, and certainly not transparent, for all comments to be confidential. It also means the author cannot directly address the points made by the reviewer.

Reviewing checklists and forms generally include questions that cover some or all of the following:

Scope. Does the paper fall within the scope of the journal? Editorial pre-screening should have picked up many of those that are not suitable (see Chapter 3, page 36), but some will have got through the initial screen, either because of omission or because it was not obvious without going through a full review.

Research objectives. Are the research objectives clearly stated? How important are they? Have the objectives been met, or is the study incomplete or too preliminary?

Study design and methodology. Has the study been correctly designed? Has the appropriate methodology been used, and has it been used correctly? Are there any flaws or omissions, and how serious are these? Have the appropriate statistical analyses been done, and have they been done correctly?

Soundness of results. Are the data sound? Are there enough repetitions of the experiments? Can all the data be accounted for?

Interpretation. Have the data been analyzed and interpreted correctly? Are there possible alternative interpretations? Have all other related studies been taken into account?

Originality and significance. Is the work original or is it rather confirmation of work done by others? How significant are the findings? Do they represent a large advance in current knowledge and understanding, or are they just an incremental advance?

Existing literature. Has the background been presented appropriately and the existing literature cited adequately? Has due credit been given to other work and has the present work been set in the correct context?

Presentation. Is the presentation clear and logical? Has the manuscript been written in a way appropriate for the journal and for the section for which it is being considered? Are there any language problems? Is the length appropriate for the content? Are the numbers of tables and figures excessive? Would some of the text be better put into tables and/or figures? Are there any nomenclature problems? Have all the supporting materials been submitted? Is all the supplementary material necessary? Should any of the supplementary material be included in the paper itself? Would some of the manuscript be better featured as supplemental material for online publication?

Policy requirements. Have all the journal's policy requirements been met? These can be listed to prompt reviewers – for example, have accession or clinical trial numbers been provided, have the authors agreed to free distribution of materials reported in the manuscript, have relevant data been deposited in a public database, have all ethical considerations been taken into account?

Suitability for the journal. Whereas most questions on reviewing checklists ask reviewers for quite objective assessments, suitability for the journal is a subjective one – but one which can be helpful to the editor, especially when it comes from an experienced and knowledgeable reviewer who is used to reviewing for the journal, or which they can choose to ignore (see Chapter 5, page 89). For those journals that request that reviewers do not include any recommendation on acceptance or rejection in their reports for authors, all comments on suitability will need to be put into the confidential parts of the reviewing form.

The *BMJ* is very active in evaluating the quality of peer review. As part of that effort it has been attempting to define what a good-quality review should contain, and has developed a 'Review Quality Instrument' (RQI), that can be used to evaluate various aspects of a peer-review report.[12] This summarizes those in eight (in v3.2) clearly defined questions covering: discussion of the importance of the research question, discussion of originality, identification of the strengths and weaknesses of the method, usefulness of comments on presentation, constructiveness of comments, substantiation of comments, comments on interpretation of results, and an overall rating of the review. Each question is evaluated on a scale of 1 (poor) to 5 (excellent). An overall measure of review quality is obtained by taking the mean of the scores for the eight questions. The RQI has a limitation, however, in that, although it can assess the quality of a reviewer's comments, it cannot assess the accuracy of those comments. The *BMJ* has a large number of checklists on its website that cover both general and specific areas (such as statistical review and technical points), and many editors will find it helpful to take a look at these (http://bmj.bmjjournals.com/).

Associated material needed by reviewers

For reviewers to be able to carry out a complete review it is important that they receive all the relevant materials and documents in addition to the manuscript itself and the usual instruction sheets, forms and checklists. These could include the following.

Supplementary material
As mentioned in Chapter 3 (page 31), supplementary material can be of two kinds: either additional material that is to be used just during review to aid assessment and is not destined for publication, or additional material that will be published only in the online version of the journal, not in the paper copy, if the paper is accepted.

Supplementary material is becoming increasingly important in certain disciplines, for example molecular biology, where enormous data sets (raw data) are being generated. It can be very important in assessing work, and reviewers should be made aware of it and asked to review it accordingly; the supplementary material is an integral part of the manuscript and should be peer reviewed. It is good practice to ask reviewers whether they feel the supplementary material is important and appropriate enough to appear with the published paper online (or perhaps be actually added to the paper itself) or whether it is superfluous and not needed. Authors will sometimes try to get into press 'leftover' material or data that aren't really appropriate to the paper under review and that they wouldn't be able to get published on their own anywhere else. Supplementary material can be anything from simple figures or additional tabulated data sets to complex figures or movies. Extra figures or text won't be a problem and can be sent to the reviewers in the same way as the manuscript. Large data sets, however, can cause problems and need special consideration. If they involve extensive spreadsheets that the reviewers need not only to view but also to analyze and manipulate, it isn't practical to send them as hard copy printouts (they may take up hundreds of pages) or as PDF (portable document format) files (they won't be readable or analyzable). They need to be sent on a disk or CD if hard copy manuscripts are still being sent. If a journal has an online reviewing system, the files can normally be uploaded separately and made available to the reviewers via the system. If a journal has an FTP (file transfer protocol) site (used for transfer of files across a network), the files can be mounted there and the reviewers given instructions on how to access them.

Increasingly, authors are providing URLs to various websites in their manuscripts. Some will be to general sites or public databases and be a source of additional external information that will be of interest or use to readers of the article, but not of direct relevance to the review of the manuscript. Others may contain information or tools that are critical to the review process. These sites need to be checked. If any are to an author's own site they may be problematical, as visitors to websites can be identified and reviewers' anonymity could therefore be compromised, which is an important consideration if a journal is operating closed peer review. The editors, editorial staff and reviewers may be the only people visiting the site, especially if it is password controlled and they have been given a user name and password as part of the submission. They could be readily identified if an author chose to set about doing this. Many authors would be appalled to be suspected of this, and wouldn't dream of doing it. But as the possibility exists and could have serious implications, it needs to be considered. Some reviewers will not be aware that they could be identified. Others will suspect that this may be the case and contact editorial staff to express their anxiety; some may refuse to access certain sites. If reference is made to an author's site and the material mounted there is important for the review process, reviewers should be alerted and the material must be made available to them by other means. Ideally, guidelines should be given in a journal's information for authors that they should avoid citing their own websites for material that requires review. They should rather submit that material online, by email, on CDs or as paper copy, as appropriate to the journal's system and for the nature of the material. Reviewers will need to be notified about how they will receive the material if it is being sent separately or by another means to the manuscript and the rest of the material. Screening manuscripts for suspect websites that are needed for review will take some specialist knowledge and may be beyond the skill of editorial staff. An editor's assistance may be needed or communication with the author to check whether material is relevant for review. Journals have a duty to alert reviewers to the potential problem, for example by adding appropriate information to the cover letter sent to them:

> Please be aware that, as a general principle, if you access any websites your IP address, and so your institution, and possibly you, can be identified. If there are any websites within this manuscript that you need to access to obtain information or materials to enable you to carry out an accurate and thorough review but that might compromise your anonymity, please let the editorial office know. We will endeavour to provide you with the information you require by other means.

Reviewers can be advised of the existence of various Internet anonymizer sites, through which they can go ('anonymous surfing') to access an author's site without their identity being compromised. However, unless journals can vouchsafe their security and reliability, they should not recommend any particular service and include a disclaimer accepting no responsibility for any anonymizer site used.

Beware!

Reviewers can be identified by their visiting authors' websites as part of the review process.

Related manuscripts in press or submitted to other journals
Sometimes authors may have manuscripts in press elsewhere or that they have submitted, or are about to submit, to other journals that are closely related to the one under consideration. It is important that reviewers have access to these so that they can carry out a complete review. The data in these other manuscripts may be important in assessing the data and/or interpretation in the manuscript under review. There may be duplication of results or enough overlap to reduce the significance or novelty of a submission to below that acceptable by a journal. Acceptance of a manuscript may even be dependent on prior acceptance or publication of another manuscript (or manuscripts) and there will be problems if the manuscript with the other journal ends up being rejected. For all these reasons, reviewers who don't have the opportunity to view related manuscripts may return incomplete reviews, saying they were not able to assess certain parts, provide feedback on all questions, or give an indication of novelty/significance because of inadequate information. Such a review will be useless to any journal striving to maintain standards and publish work

of high calibre. Journal guidelines should therefore ask authors to include related manuscripts in press or submitted elsewhere with the submission to the journal (see Chapter 3, page 30) and these should be made available to the reviewers. A note of caution: it should be made very clear to the reviewers which manuscript is the one to be reviewed and which is/are accompanying related manuscripts in press or submitted elsewhere so that reviewers don't accidentally review the wrong manuscript! It does happen and can cause enormous problems and delays if not discovered until the reviews are all in and ready for editorial assessment. For both paper copy and online review, manuscripts should be labelled appropriately: by labelling clearly on the hard copy or assigning informative file names, respectively.

Beware!

Related accompanying manuscripts should be clearly distinguishable from the manuscript intended for review to avoid reviewers assessing the wrong manuscript.

Monitoring review progress

The reviewers have agreed to review a manuscript. They've been sent the manuscript and are, hopefully, reviewing. All journals will have experience of the whole range of reviewers – from those who nearly always return their reviews within days, to those who always take longer than they agreed. There may be good reasons for reviewers taking longer than expected: they may be dealing with a complicated or problematical manuscript, there may be items missing, their personal circumstances may have changed. It's important, therefore, to strike the right balance when checking on review progress and trying to get overdue reviews submitted. Contact needs to be enough so that any problems are brought to light and sorted out in a timely way, but not so intense that reviewers feel hounded and that they are being bombarded with too many reminders. Reviewers may well start to ignore reminders and might get very irate. The sending of repeated reminders that arrive, for example, within a single day before the reviewer has even had time to read or respond to the earlier ones (which, according to personal reports, does happen) can be intensely annoying, and is far too demanding of anyone. Receiving repeated reminders can become especially annoying if a reviewer has pre-arranged with a journal that their review will take longer than usual. A note of this should be made (in the reviewer notes screen; see this chapter, page 44, and Box 4.1) and actions modified accordingly. If an online system is being used that automatically sends out programmed reminders, the reminder schedule should be adjusted; if it can't easily be changed, reviewers should be pre-warned that automatic reminders will arrive but that they should ignore them. When 'real' reminders are sent, these should be worded to show that the journal is aware that the review was going to take longer than usual but that time has now been exceeded.

Reminding and chasing reviewers

Different journals have different expectations of how long they want reviewers to take to assess manuscripts and return reviews. In fast-moving fields and for the top science journals, a period of 7 to 14 days is usual. Fast-track papers might need to be turned around even more quickly. In some disciplines, a month, or even months, may be the norm. The ALPSP/EASE online peer-review survey in 2000[6] found that the most common deadline for review was 3 to 4 weeks (the time for around 70% of the respondents). For around 15% of respondents it was under 2 weeks, for around 12% it was more than 4 weeks, and around 4% of the journals didn't have any deadlines. Reviewers will have been made aware of the deadline for their reviews when originally approached or in the review materials sent to them. The expected review time will also dictate when reviewers start being chased for their reviews, as will the mode of manuscript transfer. If paper copies are still being mailed out, mailing time (and so geographical distance away from sender) will have to be taken into account when working out a reminder schedule.

How reviewers are reminded will also depend on the peer-review system in operation. If paper copies are still being used and reminders have to be generated manually, a forward diary or equivalent needs to be kept that shows when reviews are due and those dates marked ahead. Once the expected return date is exceeded, the forward reminder schedule needs to be added. With online systems, it is usually possible to set up a reminder protocol when manuscripts are assigned to reviewers – and reminders will go out automatically at preset times until the review is returned or a last-automatic-reminder stage is reached and more drastic, customized, measures need to be taken.

Irrespective of whether a paper-based or online system is used, certain considerations are the same when sending reminders. Nowadays, email is normally used by most journals, but fax and phone can play important roles, as can personal contact if the opportunity presents itself. But many editors will have experienced tardy reviewers fleeing in the opposite direction when they see them coming (and the opposite occurring when these individuals are authors, when they will actively pursue an editor to discuss the decision on their manuscript or to complain at how long it is taking to review).

The following is a typical reminder schedule, and Box 4.2 gives examples of the sort of messages that can be sent at the various reminder stages:

1   Any agreed changes to the usual review time should be noted with the manuscript, preferably in the reviewer information notes screen (see this chapter, page 44, and Box 4.1).

2   The first 'reminder' can be a gentle message checking that the manuscript has arrived safely or that the reviewer has been able to access it without problem and inviting them to contact the editorial office if there are any problems.

Box 4.2 Examples of reviewer reminder messages for an online review system

Note: The time intervals at which the messages below are sent out will depend on the journal's normal expected review time, which the reviewer should have been made aware of at the time of agreement to review. If later return has been agreed with the reviewer, the reminder schedule should be adjusted accordingly.

First 'reminder'

'Thank you very much for agreeing to review the above manuscript for [name of journal]. We sent you details of how to gain access to the manuscript on [date]. This email is just to check that you received this information, that there are no problems, and that you have been able to view the manuscript and associated material on the journal's online submission and review site: [URL].

If you have any queries or need any assistance, please don't hesitate to contact me.

I look forward to receiving your review shortly.'

Second reminder

'Thank you very much for your help in reviewing this manuscript for [name of journal]. We sent you details of how to gain access to the manuscript on [date]. This email is just a reminder that your review is now due. So that we can provide rapid turnaround for our authors, it would be very helpful if you could submit your assessment and comments as soon as possible. Please view the manuscript and submit your review via our online submission and review site: [URL].

If you would like me to re-send the letter with instructions on how to view the manuscript and submit your review online, I would be very happy to do so.

If you will have any problem reviewing the manuscript within the next few days, please could you let me know as soon as possible.

Thank you very much for your help, and I look forward to hearing from you very soon.'

Third reminder

'Thank you very much for your help in reviewing this manuscript for [name of journal]. You were given access to the manuscript on [date] and your review was due in on [date]. We do not, however, appear to have received a review from you. As we would like to return a decision to the authors as quickly as possible, we would appreciate it if you could send us your comments without delay.

Please view the manuscript and submit your review via our online submission and review site: [URL].

If you would like me to re-send the letter with instructions on how to view the manuscript and submit your review online, I would be pleased to do so.

If there has been an unforeseen problem and you will not be able to return your review within the next few days, please could you let me know straight away.
I look forward to hearing from you soon.'

Final, customized reminder

'On [date] you agreed to review this manuscript for [name of journal] and to return it by [date]. Unfortunately, we still haven't received your review and it is now [time] overdue. We have notified the handling editor, [name of editor], who is very concerned about the situation, especially since you let us know [time] ago that you were close to completing your assessment of the manuscript. [Name of editor] has asked that you submit your review or let us know immediately if you are no longer able to carry out an assessment of the manuscript. The authors are, understandably, very anxious and unhappy that the review of their manuscript has been so delayed.

If you are having problems submitting your review via our online site, we would be very happy to receive your comments by regular email.

Please accept my apologies if this email has crossed with submission of your review. If it hasn't and we don't hear from you, I will call later today to make sure our messages or your review haven't got lost en route.'

3   The next reminder will usually be the first one sent because the review deadline has passed.

4   A number of subsequent reminders may need to be sent. These should be worded appropriately and reflect an increasing urgency. They should, however, always be courteous and never aggressive or threatening. With some online systems the same message may be used for the first few automatic reminders but then the opportunity is given to change the wording in the final, customized reminder. Care should be taken to devise an appropriate message header for reminders so that messages don't get classified as 'spam'. Also, including the journal's name makes it more likely that a reviewer will take a message seriously and read it.

5   The wording in the final, customized reminder needs to be appropriate to both the situation and the specific reviewer involved. If a reviewer has repeatedly promised to return their review and it never comes, a measure of toughness may be called for. If a reviewer has found themselves in serious personal circumstances, then a gentle approach will be more appropriate, with perhaps the suggestion that they abandon the review. We once had a case of an overdue reviewer sending a message from his hospital bed after being involved in a car accident, apologizing for the delay and very nobly promising to do the review as soon as possible. This is beyond all reasonable call of duty, and compassion is clearly called for in such circumstances.

6   Resorting to telephoning a late reviewer can work and elicit a review. But some reviewers cannot be contacted by phone or will also ignore those reminders and then more drastic measures need to be taken (see this chapter, page 72).

Beware!

Take care to avoid email headers that may classify reviewer reminder messages as 'spam'.

Problems during review

Throughout the review process, it is important that a dialogue is kept up between the editorial staff, reviewers and editor. Problems may crop up for which reviewers need help or advice, or on which editors need to be given feedback and may need to advise on action to be taken. What kinds of things might arise?

1   A reviewer finds a problem with a manuscript. If this is relatively trivial and straightforward, it can be dealt with by the editorial office. For example, a wrong figure or table may have been submitted in error by the authors, perhaps one from an earlier draft of the paper (hopefully, if all the right checks were made on submission – see Chapter 3, page 32 – there shouldn't be any omissions). The authors need to be contacted and asked to send the correct item(s) as quickly as possible to avoid too much delay in the review process. Some more serious problems may come to light, for example suspected fraud or plagiarism, and the review process may need to be halted while they are resolved (see Chapter 9, page 186).

2   A reviewer finds they will be considerably delayed in returning their review or they may no longer be able to review a manuscript. Whether another reviewer needs to be found to replace them will depend on a number of factors and should be evaluated by the editor. If the other review/s is/are in and is/are thorough and helpful, then the editor may feel there is no need to find a replacement reviewer. Conversely, if the other review/s is/are superficial or too brief to provide a thorough assessment, the editor may decide to go out for another review. The course of action will also depend on what the reviews say. If they suggest that there are serious flaws and the editor finds these to be correct, there is no need to go for another review as those flaws render the manuscript unacceptable. If the original reviewer whose review is going to be late is a particular specialist or someone whose opinion the editor values highly, the editor may decide to wait for their review even if it's going to be considerably delayed. In cases of extended review, let the authors know what is happening as this will help maintain good relations. Many authors will be glad to be kept informed and won't begin to wonder if their manuscript has been forgotten or start to suspect that someone is intentionally delaying the review process.

3   The authors contact the journal with information about a change in circum-
    stances. These will range from the trivial to the serious. Editorial staff will be able
    to deal with the former – such as address changes, missing sentences, periods
    of absence – but will need to contact the editor about more serious issues – for
    example, if the authors find in work done subsequent to submission that the
    data they have submitted are not as sound as they originally thought, or that the
    new data change the interpretation given in the submitted paper. How such
    cases are dealt with will depend on the severity or extent of the problem. It may
    be sufficient to provide the new information to the reviewers to be used in
    assessment. Alternatively, the manuscript may need to be withdrawn and resub-
    mitted at a later date with the new data, and new interpretation if relevant,
    incorporated. In other cases, authors collaborating with commercial partners
    may find they are not able, after all, to obtain the permissions from them that
    are required to comply with journal policy – for example, if a journal has a
    requirement that materials reported in its papers are made freely available and
    distribution of some materials reported in the manuscript under consideration
    is not going to be allowed by the commercial partner. The editor will need to
    decide whether the new circumstances or restrictions are acceptable or whether
    the manuscript should be withdrawn from review and declined for further
    consideration. When a manuscript is withdrawn from review by a journal, the
    authors must be given clear reasons why this is being done and be given the
    opportunity to respond if appropriate.

4   Issues concerning authorship may arise and these need to be dealt with. The sorts
    of problems that can occur are described in Chapter 8 (page 156). If they are
    serious, the review process may need to be halted until they've been resolved.
    Journals should not get involved in authorship disputes. It is up to the authors to
    resolve their authorship problems and disagreements. Authors may, however,
    contact a journal or editor for advice on, for example, what qualifies someone to
    be an author or the order in which names should appear, or for guidelines and
    references they can cite to someone who is perhaps trying to be an author on
    their paper but who has made very little or no contribution to the work reported
    in it (see Chapter 8, page 151, for information on this).

5   If one of the authors dies during the review process (or has died quite recently)
    some reviewers may find it difficult to submit an unbiased review, particularly a
    negative review, and especially if they had personal or professional links with
    the author or admired and respected them and their scientific work. If any such
    reviewers contact a journal, new reviewers should be found and appropriate
    messages of sympathy should be sent to the original reviewer(s).

Reviewers not returning reviews

There will be times when, even in the best-run and most efficient journals, reviewers
do not return their reviews. It seems that whatever the editorial staff and/or editor

do, no review is forthcoming. If a journal didn't seek a reviewer's agreement to review before sending the manuscript out, it will have little power to put pressure on that reviewer. But if it did get the reviewer's prior agreement, it has every right to pursue them for their review or for an explanation as to why a review is not being returned.

There will be reviewers who reply to reminders and repeatedly promise to submit their reviews by a certain date and then repeatedly fail to do so. The delay may be due to sheer inefficiency and lack of commitment on the part of the reviewer, or to lack of time because of the ever-increasing workloads of academics. There will be a whole spectrum of reasons from inefficiency or heavy workload to the more sinister cases, where a reviewer may deliberately be trying to hold up the review of a manuscript. The repeated failure to return a promised review is the worst-case scenario, as editorial staff will genuinely believe for some time that a review is on its way and tend to be relieved and take no further action. It will, however, be necessary at a certain point to decide whether to abandon that review. The editor will also have to decide whether a substitute reviewer needs to be found. This will depend on the reviews already received (as described above on page 70).

Other reviewers don't reply to email reminders and can't be contacted by phone. This may be because they are away or ill. It's useful to phone the departmental secretary at their place of work to check on this. Such action can bring a remarkably fast return of review when a reviewer realizes their presence has been confirmed. If it doesn't, then the review may need to be abandoned.

There may be reviewers who also have manuscripts submitted to the journal. Occasionally, human nature being what it is, reviewers will refuse to return their review until they get the decision on their own manuscript. They may feel particularly aggrieved if it is clear from the reference numbers that their own manuscript was submitted some time before the one they are reviewing. It can turn into a deadlocked situation if the reviewer of one manuscript is the author of a manuscript that is being reviewed by the person who is the author on the manuscript they are reviewing! The reviewers need to be handled carefully, but firmly, and it made clear to them that this isn't really appropriate behaviour. Appealing to people's better natures often works.

Information on all cases of delayed or non-returned reviews should be added to the journal's reviewer database for future reference. Reviewers who regularly fail to return reviews, or who do so only after unacceptably long times despite reminders, should not be asked to review in future.

Beware!

Reviewers initiating a bargaining situation for return of their reviews.

When it proves impossible to get a final review, some editors may write a 'reviewer' report and treat it as though it had been submitted by an independent reviewer. There will be disagreement as to whether this is acceptable. Editors do it because it makes things easier. However, it isn't a particularly honest approach and is not transparent. It is far better for the editor to complete a full review of the manuscript (if their expertise is in the subject area of the manuscript and so they are able to do this) and to tell the authors they have done this.

Receiving and checking of returned reviews

Reviews for manuscripts will be being returned at different times and possibly by a variety of means – online, by email, by fax, by regular mail. All the reviews for a manuscript submitted relatively recently may be received before those for a much earlier manuscript. Irrespective of how early or late they are, the same steps need to be carried out when they are received.

The receipt of all reviews needs to be logged. Many online systems will automatically register receipt and usually send a thank-you message (with whatever else the journal would like the message to say) automatically to each reviewer. They will also send automatic notification of review receipt to whoever is indicated (editorial staff, handling editor and editor-in-chief, for example – but see Chapter 6, page 127, for things to consider when deciding about this). If a manual entry system is used, the reference number and date need to be logged, along with who the review is from. It may be too time-consuming for offices without online systems to send a thank-you message to reviewers every time a review is received. They may prefer to wait until review outcome is known and then also include the decision and reviewers' comments (see Chapter 7, page 139).

Reviews will be submitted with varying degrees of 'completeness'. Most journals would ideally like all reviewers to fill in their standard reviewing forms in full as well as submitting their free narrative comments (see this chapter, page 61). However, circumstances may arise when a reviewer is not able to access a journal's reviewing forms, or they may not have them with them, for example if they're travelling. If that reviewer's review is not the last to come in, the reviewer can be advised to wait and submit it when they are able to access the forms. However, if a reviewer is keen to get their review finalized and submitted then it would be foolhardy not to accept it as it is and make the necessary adjustments (such as submitting it via their online account for them, brushing up the language, or expanding abbreviations), otherwise they may forget to submit it later or mislay the file or paper copy, and the review will be lost to the journal. If that reviewer's review is the last one outstanding, and especially if it is overdue, most journals will be keen to accept the review in any form. They may be desperate for it! Occasionally, a review may, due to unavoidable circumstances, have to be given verbally. In these cases, the completeness and accuracy of the written report produced as a result will depend on the ability of the person receiving the report and their confidence in being able to take reliable notes. A simple assessment for a straightforward manuscript without too much technical detail will be easy for most editorial staff. A lengthy review for a difficult manuscript with a lot of technical detail will be virtually impossible for a non-specialist to relay unless it is dictated. In the latter case, it is better to get a skeleton review with opinions and recommendations verbally, enough possibly to enable the editor to make a decision (although he or she may decide, depending on the other reviews and information available, that it is necessary to wait for the full review), and ask the reviewer to submit a more detailed one as soon as possible.

As well as making sure reviews are logged accurately – this is crucial so review status of manuscripts can be tracked – it is good practice to run certain checks on all reviews received. These checks can be done on screen if reviews are being submitted online or by email, but many people find it easier to print them out to do this. They can then be kept in a hard copy file for future reference and will provide a useful backup if the network or online system is down. Paper reviews are also very helpful when dealing with related or similar manuscripts, as it is easy to compare them side by side. It is good practice to check reviews as they come in rather than waiting until they are all in for a particular manuscript. There may be errors that need to be corrected, clarification may need to be obtained from the reviewer, or additional action may need to be taken based on the comments or recommendation of that reviewer. This can all be being sorted out while the review process is going on and will prevent or minimize delays.

Checking of reviews

As mentioned immediately above, reviews should be checked as they are received. What checks need to be made?

Reviewer anonymity is not compromised

Reviews may be submitted in a variety of ways and if a journal operates a closed peer review system (see this chapter, page 41), checks will need to be made to ensure that reviewer anonymity is not compromised (see Golden Rule 4). If reviews are faxed, header identification needs to be removed before they are photocopied if a paper-based system is still used or before they are scanned to create digital files. If reviews are submitted by email as attached files, it has to be remembered that the creator of those files can be identified if the creator hasn't removed their identifying details, and so this will need to be dealt with, for example by creating a PDF of the report meant for the authors or by using the 'Remove Hidden Data' or 'Remove Personal Information' tool available with certain software (and usually found under the security and privacy options). If a reviewer has submitted a marked-up PDF of the manuscript, particular care must be taken not to leave any identifiers as the creator's name may be attached to every note or change made.

Beware!

The creator's identity may be attached to electronic files, and in a number of places – make sure all cases are removed before files are transferred to authors unless open peer review is used.

The correct review has been submitted

The manuscript reference number should be checked against the manuscript title and authors' names if a reviewer has entered these themselves, to make sure they agree. A reviewer may be entering them because they are not using the standard reviewing forms but are creating a document to submit, or because a journal's standard reviewing forms are, by design, blank when a reviewer accesses them. Many journals do enter these details, and with most online systems it happens automatically. In both these cases, the reviews should be scanned visually to make sure the content of the report does refer to the current manuscript. It is not unknown for reviewers to pick up the wrong files from their computers, or to start filling in the wrong reviewing form. Reviewers who are reviewing more than one manuscript at a time for the same journal are particularly at risk. It can be difficult in some cases for a non-specialist or non-scientist to pick up this kind of error, but in many cases it will be obvious: the reviewer may include a manuscript title or the authors' names in their actual report and these may be different from the ones for the review they are supposedly submitting; the terminology or substance may seem very different from the title or author's cover letter. Picking up such an error at this stage can save a good deal of trouble and delays later on, as the reviewer needs to be contacted and the correct report obtained. If an online system is being used, the service provider may need to be contacted to make the technical adjustment to the database to allow a reviewer's report to be changed for a manuscript if the system doesn't allow editorial staff to do this. It can also save potential embarrassment if an editor is not perhaps as observant as they should be and makes a decision (either final or to send the manuscript out for further review) based on the reviews received without noticing that a reviewer has submitted the wrong review or report, and it is only when the author receives the decision and reports that the error is brought to light. To let an error get this far does not reflect well on a journal and indicates a certain lack of care in editorial processing and decision making. It is the duty of every journal to ensure that the correct reviews and reports have been submitted before the editorial decision is made. Extra-careful checks need to be made in cases where a reviewer has accompanying related manuscripts included with their reviewing materials in case they have, despite having been given clear guidance, reviewed the wrong manuscript (see this chapter, page 66).

Beware!

Watch out for the wrong review or report being submitted for a manuscript in error. Reviewers who are reviewing more than one manuscript for a journal or have an accompanying related manuscript to review are particularly at risk.

Comments have been entered in the correct places

Many journals have forms that have sections for confidential comments to the editor as well as for those comments destined to go to the author (see this chapter, page 61). A quick scan will show if a reviewer has inadvertently put them in the wrong place. This will need to be corrected and the reviewer contacted if their input is needed to provide clarification or additional comments. For hard copy forms, a bit of cutting, pasting and photocopying will usually work. For online systems, the service provider may need to be contacted about this. In these systems, 'Reports for Authors' are usually picked up automatically when a decision is sent to the author, and so it is essential that comments are moved to the correct position before the decision is sent or the author will see the comments that the reviewer thought he or she was making in confidence to the editor. An alert note needs to be added to the notes or notes screen for that manuscript and the editor alerted to hold on submitting their decision. Sometimes reviewers will only put comments in the section for confidential comments for the editor. A quick read will determine whether the reviewer has probably intended these for the authors and the comments can be moved. It is good practice to add a note to the manuscript information to indicate that this has been done without needing to check with the reviewer and who has done it. If there is any doubt as to whether the comments can be sent to the author, the reviewer should be contacted to check on this and to get further comments if the ones submitted were intended solely for the editor. Reviewers should provide comments for the authors (see this chapter, page 62). Authors expect to see them as a product of the peer-review process and can get annoyed or suspicious if they don't get any. However, if reviewers insist that they will provide only comments for the editor and will not move from this position, these will need to be incorporated into the editor's decision letter or compiled into a suitable report. As this can involve a significant amount of time and effort in some cases, this is another reason reviewers should be encouraged as strongly as possible to submit comments that can be sent to the authors.

There are no unprofessional or derogatory comments

Reviewers should review manuscripts objectively and not make personal or derogatory comments about the authors; they are reviewing the work, not the people involved (see Golden Rule 6). Reports should be skimmed to check for any such comments. If any are found, the reports will need to be edited (see this chapter, page 77).

Unusual cases requiring immediate action

Occasionally, unusual situations crop up that need immediate action. Reviewers may suggest in their reports that person X should definitely be asked to give an opinion on the manuscript or that a certain type of specialist needs to assess it. It would cause unnecessary delay to wait until the end of the review process to act on this, and so the editor or person dealing editorially with the manuscript should be alerted to this information. They may well request that an additional reviewer be contacted or that the other reviewer(s) be provided with some additional information. Occasionally (although from personal experience it appears to be becoming more common – or perhaps it is just being picked up more often), a reviewer may make an allegation of fraud or plagiarism against the author. Some reviewers will be hesitant about even mentioning the possibility of wrongdoing, others will be very adamant and vigorous in their demands that something be done. These are very serious accusations and must be investigated (see Chapter 9, page 184). They may be time-consuming and difficult to deal with, but editors are ethically obliged to look into them.

The ethics of amending reviewer reports for authors

Opinions amongst journal editors will be divided on whether it is acceptable to amend reviewer reports for authors. There will likely be little disagreement that any potentially libellous or derogatory comments should be edited out. But, beyond that, practice will vary and, unfortunately, some editors may choose to make their lives easier by selective editing of reports so that they better reflect their editorial decision. Reports meant for the authors should not be edited in this way. Not only is it unethical – why then have a peer-review process? – it can lead to unpleasant situations and angry authors and reviewers. Reviewers may reveal their identity to the authors, perhaps if they meet up at a conference, and be surprised to hear that their comments were not passed on accurately to the authors, or maybe not at all, and their opinions misrepresented. It is good practice to let authors have all the reviewers' reports for authors and for editors to explain in their decision letters if they are overriding anything a reviewer has said, or that they don't agree with, and to give reasons for this (see Chapter 5, page 89). Editors should aim for their decision making to be as transparent as possible and they should be able to substantiate their decisions if challenged.

What sorts of things might need to be changed in a report intended for the author? The following are based on real examples.

Derogatory or potentially libellous comments

> X has never had an original thought in his life and wouldn't know what to do with it if he did.

Derogatory comments like this should be removed. Not only is this a totally unacceptable comment to make, it suggests that the reviewer may have been assessing the person, not the work, which is wrong (see Golden Rule 6).

Unfortunate use of language
Occasionally, reviewers whose first language is not that in which they are writing the report will use words that are not quite right. Most of the time they can be left; authors appreciate that reviewers come from an international community and so expect there to be general language errors and they make allowances for these. However, if usage is such that it might cause offence or imply something the reviewer clearly didn't intend, changes should be made. Take, for example, the following comment:

> I think an honest scientist would calibrate the two instruments used in the study and recalculate the relative activity values.

It was clear from the rest of the report that the reviewer hadn't intended to accuse the author of being dishonest, but had merely used an unfortunate choice of words in his report. The report was amended to:

> The two instruments used in the study should have been calibrated and the relative activity values recalculated.

Beware!

Watch out for derogatory comments or unfortunate use of language in reviewers' reports for authors.

Use of colloquial language

English is the predominant language of scientific publishing and so the majority of scientists will review and write reports on most manuscripts in English. The authors for most journals will come from many countries around the world. Not only will their ability to write in good English vary, their ability to grasp meaning will also be very different. Although they will be familiar with commonplace words and use of language, they may well not understand colloquialisms or country-specific phrases. These should, therefore, generally be avoided, and if unusual or strange examples are spotted in reviewers' reports it is doing a service to the authors to change or delete them. For example, the following might have left the Chinese authors struggling to understand what was meant:

> Persistence is admirable, but throw the ball over the plate with the next pitch.
> Fercryinoutloud, ... [This 'word' wouldn't appear in any dictionary!]

The 'ideal' report

All journals will have experienced receiving the whole range of quality of report, from very abrupt two-line comments to lengthy contributions that cover many pages. Although it must always be remembered that length does not reflect quality, in general a very brief report is not ideal and won't be very helpful: it's very hard to tell whether the reviewer has actually read the manuscript properly and it is, for example, a really great piece of work that can be published virtually as it stands, or whether they have read it only superficially and that's the reason why their report is so brief as they can't make any constructive comments. Likewise, lengthy reports are not always helpful in aiding editorial decision-making. The reviewer may have listed numerous style or language points, but completely failed to provide a proper assessment of the work.

If a manuscript is really bad, reports may be either very long or very short. Once the reviewers realize that the quality is very poor, they may just list the major flaws and not spend the hours that would be needed to list everything that is wrong with the manuscript. But some reviewers may, in the spirit of collegiality and helpfulness, do so, especially for manuscripts that come from developing nations or from groups they feel deserve their help. Although this will undoubtedly be of great value to those authors, it is beyond what can reasonably be expected of reviewers. A note of caution about manuscripts that are too preliminary or very complex: I have been told on two separate occasions by different people that they know of groups who have submitted manuscripts in the full expectation that they would be rejected – they just wanted to get the reviewers' reports to find out about any deficiencies and to get ideas on what they needed to do and suggestions for experiments and how to restructure the manuscripts. This may be a flattering comment on the perceived quality of the journal's review process, but it isn't really honest behaviour on the part of the authors.

It is a good idea for editors and editorial staff to be able to describe what constitutes a 'good' report so that feedback can be given to reviewers if it's clear they need some general guidance, or advice can be given to new, inexperienced reviewers who may have reviewed very little and not had enough papers published to have seen many reports done by others. Reports should generally be structured and include certain information (see Box 4.3). As mentioned before (this chapter, page 61), it's also very important that what reviewers write in their narrative reports for the authors corresponds to what they have written in the confidential reports and forms submitted for editorial use and the evaluations and recommendations they have made there. Reviewers may also require some guidance on what sort of things they should put into their confidential comments to the editor (see Box 4.4). This section should contain only those things that genuinely need to be confidential as, in the interest of transparency, everything that can go into the report for the authors should appear there. All reviewers should bear in mind the following points when they are preparing their reports:

- Write clearly and in a way that will be easily understood by people whose first language is different from the one in which you are writing.
- Avoid complex or unusual words, especially ones that would cause even native speakers to resort to their dictionaries.
- Be objective, not subjective.
- Be constructive, not destructive.
- Substantiate any serious charges: for example, of prior or duplicate publication, fraud, plagiarism, unattributed work.
- Treat the author's manuscript and work as you would like your own to be treated.
- Number your points and refer to page and line numbers in the manuscript when making specific comments.

Box 4.3 Advice to reviewers on how to prepare an 'ideal' report for authors

1. Summary

- Start with a brief summary of what the paper is about and what the findings are.
- Put the findings into the context of the existing literature and current knowledge.
- Indicate the overall significance of the work and whether it is novel or mainly confirmatory.
- Give an idea of the quality and completeness of the work; indicate its strengths.
- State whether there are any major flaws or weaknesses.
- Note any special considerations – for example if previously held theories are being overturned.

2. Major issues

- Are there any flaws (technological, design, or interpretation), what are they, and what is the severity of their impact on the findings?
- Has similar work already been published without the authors acknowledging this and how does the current study relate to the published study (or studies)? Does it present similar results that reinforce any other studies, or results that contradict them?
- If the authors are presenting findings that contradict current thinking, have they presented strong enough evidence to substantiate their case? If not, what additional data would be required? Have they cited all the relevant work that would contradict their thinking and addressed it appropriately?
- If major revisions are required, what are they?
- Are there major presentational problems? What are they? Are they serious enough to prevent you carrying out an accurate assessment of the work or to prevent readers understanding it? Are the problems related to language, manuscript structure, or data presentation?
- Are there any ethical issues? If there are, what are they?

3. More minor issues

- Are there any places where meaning is unclear or ambiguous? How can this be corrected?
- Are the correct references cited? If not, which should be cited instead?
- Is citation adequate to reflect other work? Is it excessive, limited, or biased?

- Are there any factual errors? What are these?
- Are there any numerical or unit errors? What are these?
- Are the figures/diagrams/plates/tables appropriate, sufficient, and properly labelled? If not, indicate which are not.

4. Expansion of questions/comments made in the reviewing checklist

These will vary from journal to journal and will depend on what questions are on their forms. Examples are:

- Do the keywords accurately reflect the content? If not, suggest alternatives.
- Are there any nomenclature issues? If there are, what are they and how should they be corrected?
- Have all journal policy requirements been followed? If not, which have not?
- Have all appropriate depositions of data and materials been made and reference numbers provided? If not, give details of omissions.

5. Opinion

Briefly summarize your opinion of the work (but note that some journals explicitly request that reviewers do not make any recommendation as regards publication in their report for the authors; see this chapter, page 61).

Box 4.4 Advice to reviewers on preparing confidential comments for the editor

- Do not use this section as a place for backstabbing. Authors will not see the comments direct and so will not be able to respond to any negative comments or opinions. They will not be able to present their side of the story unless asked to do so by the editor.
- Do you have any concerns about misconduct regarding the work or the submission? Give details.
- Do you have any additional information that might help the editor but that you cannot put into the report for authors, for example because it might identify you?
- Do you feel you are not able to assess adequately all parts of the manuscript? Give details. Are the opinions of any other specialists required?
- If you have any potential conflicts of interest, give details.
- If anyone helped you with the review, give their name.

■ Don't write all your comments on the manuscript itself instead of compiling a report. Many journals will be very happy to receive copy marked up for improvement in addition to the regular report to pass on to the authors, who will often be very grateful for a reviewer's input (but some journals will not be and may state so in their guidance for reviewers, so check first). If you do return a marked-up hard copy of the manuscript for the author, check that no other material has slipped between the pages. If you return a marked-up PDF or other electronic version of the manuscript, remove all identifiers if you want your identity to remain anonymous.

Back-to-back manuscripts

Sometimes papers are published 'back-to-back' in a journal, i.e. one after the other. In some cases the authors, either the same group or two or more different groups, may intentionally submit related or companion manuscripts and request consideration for joint publication. In other cases, different groups may unknowingly submit similar manuscripts at around the same time to the same journal. In both cases, the manuscripts need to be considered together. They should therefore be handled by the same editor or by editors working closely together.

The manuscripts should also be assessed either by the same reviewers or by some of the same reviewers. The former applies if the subject matter and content are fairly similar; the latter if the manuscripts overlap but have additional content that requires assessment by different specialists. All the reviewers should, however, have access to both the manuscripts (or however many are involved) in order to be able to carry out a complete review. Reviewing two or more manuscripts imposes a considerable burden on reviewers and so they should be sent special messages when being invited to review companion manuscripts, explaining the situation and how important it is that the manuscripts be reviewed together. Reviewers should also be given longer than usual to review (remember to note this in the notes or notes screen for the manuscripts and to change the reminder schedule).

Certain problems can arise with manuscripts considered for back-to-back publication:

1   If both manuscripts are either from the same group or from different groups who have agreed to submit simultaneously, one manuscript may be much stronger than the other. The authors may or may not be aware of this. In some cases they may be trying to get a weaker or less complete manuscript published on the back of the merits of the other. This must not be allowed to happen, and the decision on each manuscript must be appropriate and in line with the journal's usual standards. If one manuscript is not accepted and no resubmission is invited, it is important that if the other one is accepted it can stand alone and that the results and interpretation are not dependent on the prior publication of the other manuscript. If the two are very closely linked and interdependent, the

solution may be to invite the authors to combine the manuscripts, particularly if there is a lot of duplication of methodology. Authors may in some cases be reluctant to do this because they end up with only a single paper, especially if it is important for any graduate students or post-doctoral fellows to have separate publications to their names. They may be concerned that their contributions appear to be diluted if the number of authors increases considerably. Also, there can only be one 'first' author on the merged manuscripts, whereas before two authors enjoyed that position. Part of the last problem can be alleviated by adding a footnote to the published paper indicating that the first two authors contributed equally to the work and should be considered joint first authors. In many cases, manuscripts can be combined very successfully and a much stronger and more complete paper results.

2 If two related manuscripts are from different groups of authors who are each unaware of the other manuscript, those authors cannot be told of the existence of each other's manuscript because the submission of a manuscript and all the details associated with it must be kept confidential (Golden Rule 3). If both manuscripts end up being rejected, or one accepted and one rejected, this is relatively straightforward. In both cases, neither group is told anything about the other manuscript. If both are either accepted or provisionally accepted subject to revision, this becomes a bit more problematical. If both require no or relatively minor revision, a good way forward is to tell each group about the other manuscript, ask them to get in touch with one another, liaise on revision, and reference each other's papers. If one manuscript requires significantly more revision than the other, and acceptance is not certain because those authors may not be able to meet all the revision requirements, then this is more difficult and each editor needs to decide whether to notify the authors or not. It may be fairest to let the revised manuscript of the paper requiring less revision come in and enter the production process after acceptance, then alert the other group and ask them to refer to and cite the accepted paper in their own before submitting their revised manuscript. It is very difficult to stipulate exactly what editors should do, because so much depends on the individual circumstances. Fairness should be the guiding principle, along with trying to achieve the best outcome for the journal and the scientific record.

3 Care must be taken to make sure that related papers published back-to-back refer to each other and that they do so in the same way; for example, they may do this by mentioning the other paper in the text and include a full citation in the reference list, or they may refer to one another in notes added at proof stage. If the two groups have been liaising, they should have done things the same way, but this doesn't always happen, and so cross-referencing needs to be checked before the manuscripts are sent off for preparation for publication. The order of publication of the papers also needs to be decided. If there is no clear content-related reason why one paper should appear before the other, then the order should be decided according to the journal's usual policy, for example in order of date of acceptance or, if those are the same, by date of submission.

Dealing with enquiries on manuscript status

Journals and editors will receive many enquiries on manuscript status. Information on manuscripts submitted to a journal is confidential (see Golden Rule 3) and should be released only to the appropriate people, i.e. the authors listed on the manuscript. Email enquiries will show whether an enquiry is being made from an author, but telephone ones won't. It must therefore first be established who is calling and that they know the details of the manuscript in question, such as the title and authors (many genuine authors won't remember the reference number so that's not a good discriminator). If the caller is not an author, no information should be released to him or her unless the corresponding author has given explicit permission for this, even if the caller appears to have quite extensive knowledge about the submission. Care should be taken if decisions need to be faxed to authors via non-private fax machines. Authors should be advised when the fax is being sent if there are any concerns about privacy, so they can make appropriate arrangements for the message to be retrieved. A cover sheet marked 'confidential' should always be used.

Some enquiries may come from people 'fishing', trying for various reasons to find out where a certain manuscript has been submitted (see Chapter 8, page 157). No information on whether such a manuscript is with a journal can be given because, as mentioned before, the actual submission of manuscripts is confidential (Golden Rule 3; Chapter 2, page 14). Situations where authors appear to be emailing from someone else's email address need to be handled carefully, as, in theory, the message could be coming from anyone. One solution is to reply that the answer to the enquiry will be sent to that author's own email address.

Editorial staff may be unsure what to do if someone calls who says they are the head of the department or institute in which the authors work. Even if they are, no information should be released, including about whether a submission has actually been made; the caller should just be advised that it is journal policy that no information can be released to anyone but the authors. Not all authors will want their heads of department to know the status of their papers, particularly if they are going to be rejected. Journals should not get involved in departmental or institutional politics.

Genuine enquiries from authors need to be dealt with appropriately and it is important that a full record is kept of all enquiries, and the responses given to the authors, to ensure accuracy and consistency are maintained. Editors and editorial staff may sometimes feel tempted to blur the truth a little to avoid an author's wrath – for example, if a review is long overdue or if all reviewers haven't yet been found for a manuscript even though it was submitted some time ago. It is always better to be truthful, and to explain if there are difficulties and how they are being dealt with. Authors will very readily pick up on inconsistencies in responses and, quite rightly, may protest or get angry if they are told different things at different times by different people (or, even worse, by the same person). They will lose confidence in a journal if they are told one day that the reviews and decision on their manuscript are in and then another day that the journal is still waiting for a review to be submitted.

> ## Beware!
>
> Don't misinform authors about the status of their manuscripts in the review process – be truthful.

Authors should not be told the decision on their manuscript informally by editorial staff based on an editor's initial recommendation until the decision has been finalized. There will be times when an editor's recommendation is modified or changed to bring it into line and make it consistent with decisions being made on other manuscripts under consideration. Also, the decision may not be a straightforward one, and staff may not be able to appreciate this because of lack of specialist knowledge or due to inexperience. An author who is told that his or her manuscript has been provisionally accepted only to find later that the manuscript is being rejected will not be a very happy one, and will undoubtedly start up a correspondence contesting the decision, which may prove embarrassing as well as time-consuming. The journal's credibility may be damaged. The opposite situation may also occur, where an author is told his or her manuscript is going to be rejected but then it is decided, after editorial consideration, that it should be provisionally accepted. The author may, in all honesty, go ahead after the informal notification and submit the manuscript to another journal. The manuscript will then effectively be under consideration by two separate journals, which is not ethical, and may lead to some unpleasant exchanges once the real situation becomes clear.

> ## Beware!
>
> Don't release editorial outcome decisions to authors prematurely; they may be wrong. Wait until they've been finalized.

References

1   van Rooyen, S., Godlee, F., Evans, S., Smith, R. and Black, N. (1998). Effect of blinding and unmasking on the quality of peer review: a randomised trial. *JAMA*, **280**, 234–237.
2   Justice, A. C., Cho, M. K., Winker, M. A., Berlin, J. A. and Rennie, D. (1998). Does masking author identity improve peer review quality? A randomized controlled trial. *JAMA*, **280**, 240–242.
3   van Rooyen, S., Godlee, F., Evans, S., Black, N. and Smith, R. (1999). Effect of open peer review on quality of reviews and on reviewers' recommendations: a randomised trial. *BMJ*, **318**, 23–27.
4   Godlee, F., Gale, C. R. and Martyn, C. N. (1998). Effect on the quality of peer review of blinding reviewers and asking them to sign their reports: a randomized controlled trial. *JAMA*, **280**, 237–240.
5   Godlee, F. (2002). Making reviewers visible: openness, accountability, and credit. *JAMA*, **287**, 2762–2765.

6   ALPSP/EASE. (2000). Current Practice in Peer Review: results of a survey conducted during Oct/Nov 2000. www.alpsp.org/publications/peerev.pdf.

7   Hartley, J. and Kostoff, R. N. (2003). How useful are 'key words' in scientific journals? *Journal of Information Science*, **29**, 433–438.

8   Schroter, S., Tite, L., Hutchings, A. and Black, N. (2006). Differences in review quality and recommendations for publication between peer reviewers suggested by authors or by editors. *JAMA*, **295**, 314–317.

9   Wager, E., Parkin, E. C. and Tamber, P. S. (2005). Are reviewers suggested by authors as good as those chosen by editors? Results of a rater-blinded, retrospective study. Abstract published in the program of the Fifth International Congress on Peer Review and Biomedical Publication, Chicago, Illinois, September 2005. www.ama-assn.org/public/peer/program.html.

10   Goldsmith, L. A., Blalock, E., Bobkova, H. and Hall, R. P. (2005). Effect of authors' suggestions concerning reviewers on manuscript acceptance. Abstract published in the program of the Fifth International Congress on Peer Review and Biomedical Publication, Chicago, Illinois, September 2005. www.ama-assn.org/public/peer/program.html.

11   ALPSP and Kaufman-Wills Group. (2005). The facts about open access. A study of the financial and non-financial effects of alternative business models for scholarly journals. www.alpsp.org/publications/pub11.htm.

12   van Rooyen, S. (2001). The evaluation of peer-review quality. *Learned Publishing*, **14**, 85–91.

5 The decision-making process for reviewed manuscripts

The organizational structure for decision making

At the end of the peer-review process for manuscripts that passed the initial editorial screening (see Chapter 3, page 36) and were sent out for external review (see Chapter 4), all the reviewers' reports, ancillary comments and any other relevant information need to be assessed by the person or people who are responsible for making editorial decisions. The organizational structure of this will differ from journal to journal. It will depend on various practical considerations aside from personal preference, such as size of journal and geographical location of the editors. For small journals with relatively low numbers of submissions, a single editor may be responsible for all decisions, seeking additional advice or opinions as and when necessary (see Table 5.1, a). For large journals this won't be practical. Most academic editors carry out editorial work on top of their research, teaching and administrative commitments and won't have the time to do this, or to do it thoroughly enough to ensure that all manuscripts are adequately and appropriately dealt with and in a reasonable time. For broad-spectrum journals, they also won't have the necessary expertise.

In large journals, editorial decision-making will often be a joint activity involving specialist editors (alone or in consultation with other editors), who are knowledgeable in the areas of research covered in the manuscripts received by those journals, and an editor with higher-level decision-making responsibility (see Table 5.1, b). The specialist editors make recommendations based on the external reviews and their

Table 5.1 Some basic schemes for the organizational structure of decision making.

Decision or recommendation made post-review by	Decision approved/finalized by	Decision communicated by
a editor A	editor A	editor A or admin. support
b specialist editor B or editor B in consultation with other editor(s)	editor C*	editor C*, editor B, or admin. support
c specialist editor B or editor B in consultation with other editor(s)	group of editors	editor B or admin. support
d specialist editor B or editor B in consultation with other editor(s)	group of editors followed by a more senior editor*	editor B or admin. support
e specialist editors or individual committee members	committee (e.g. 'hanging committee')	specialist editor or admin. support

* Editor-in-chief, executive editor, managing editor or whatever title is used for this role.

own reading of the manuscripts. They then submit these recommendations, and often a draft letter or comments for the authors, to the person (editor-in-chief, executive editor, managing editor or similar – the titles will vary from journal to journal, as described in Chapter 2, page 11) who is responsible for finalizing all decisions and overseeing the communication of decisions to authors. Alternatively, decision approval and finalization may be in the hands of a small group of editors who receive recommendations from specialist editors (see Table 5.1, c). There may even be a second step, where the editors in the group, perhaps each responsible for a section, refer their recommendations to a more senior editor for final approval (see Table 5.1, d). In some journals, decision making may be the responsibility of a committee (see Table 5.1, e) that meets regularly to discuss papers (after assessment and recommendation post-review either by specialist editors or by individual members of the committee) and decide their editorial fate – for example, the 'hanging committee' of the *BMJ* (*British Medical Journal*), the term being derived from the hanging committee used for the selection of artwork for the UK Royal Academy's Summer Exhibition. For most international journals, where editors are spread around the globe, this is not really a practical option, although virtual 'meetings' via conference calls or email discussion groups are now a possibility and worth considering. The introduction of online systems and the ability to transfer even very large files easily and rapidly by email make this feasible, because manuscripts and all relevant documents can very easily be shared. Whatever the decision-making structure of a journal, care must be taken that all decisions are made or approved by an editor and that the responsibility does not default to scientifically unqualified editorial assistants. Procedures should be put in place for busy times when systems are stretched, or when editors are away, to avoid this happening.

In some journals, the specialist editors are responsible for actually communicating decisions to authors as well as making them (see Table 5.1). This may be after consultation with other editors or on their own. However, with the latter there is the danger that decision making may not be consistent across all the editors. Regular circulation of reports of manuscripts accepted and rejected (both without review and after review) can help achieve consistency (see this chapter, page 101, for advice on achieving consistency in decision making). Responsibility for communicating decisions also adds a considerable administrative load to an editor and they may not routinely remember to send all the appropriate information, spot all the formatting and other problems, or include all the enclosures required (see this chapter, page 96). In other journals, decision making and communication of those decisions are carried out according to geographical location, with main co-editors taking responsibility in those locations for manuscripts submitted from certain areas of the world. Again, this can lead to inconsistency in decision making for manuscripts in similar subject areas unless good reporting and good communication channels are maintained between the various offices.

As mentioned previously (see Chapter 2, page 11), there isn't any consistency in what people involved in journal work are called, and this includes the people who have decision-making powers on manuscripts. There are numerous different terms used for this group of people, for example editorial committee, editorial review group, executive board, editorial board, editorial team, editorial management board, and these may be made up of greatly varying numbers of senior editors, deputy editors, associate editors, subject editors, advisory editors, executive editors in addition to the editor-in-chief. The permutations are enormous!

The decision-making process

Reviewers advise and make recommendations, editors (or their equivalent) make the decisions (see Golden Rule 5). If editors do nothing more than always go along with reviewers' recommendations on whether to publish or not, and with all their suggested revisions if a manuscript is thought to be suitable for publication, they are not doing their jobs properly. The recommendations are important, especially when made by an expert reviewer who is both knowledgeable about the work and familiar with the journal and the scope and quality of papers it is looking to publish. However, editors are privy to much more information than the reviewers, some of which it would not be appropriate to pass on to them, and they need to consider all of this to come to a decision that is both fair to the authors and appropriate for the journal. They may override a reviewer's overall recommendation or certain points in their reports. If they do, it is important that they provide the author with reasons why they have done so. However, an editor should not, apart from in exceptional circumstances, go against all the reviewers' recommendations and opinions – that is effectively dispensing with peer review. Exceptional circumstances would be if an editor picked up serious flaws the reviewers had missed, or if an ethical issue or misconduct became apparent during review, which would take priority over the scientific merits of the manuscript.

Decision making is more challenging when controversial areas or innovative new research are being considered. Sometimes, editors may take a gamble and decide that, based on the evidence available, publication is warranted even though some doubts remain. It is part of their job to be visionary. But, they are also in a position of great trust (and power) and so should never be reckless or make decisions on just a whim, and certainly not with any kind of bias. The threshold for acceptance is often set higher for controversial work or results that oppose existing theories. The onus is on the authors to convince the editor of the validity of their proposals and work. Such innovative papers will stimulate research by many groups, who will produce results that either confirm and reinforce the work or cast doubt on it. This is how science progresses and how new fields are opened up, to then either flourish or wither away. Editors should not, however, allow papers to be published with overstated claims and interpretations, especially controversial ones or those in areas that might have a significant effect on the public's reaction to, for example, health policy and the subsequent action of individuals. Some authors may be tempted to include such overstated claims and interpretations in their submissions, especially when they are trying to get work into the highest-impact journals, but they need to be moderated and must accurately reflect what the results presented show. Because the bar is set higher for controversial work, it is inevitable that some important papers will be rejected when first submitted. This has even happened for work by people who have gone on to become Nobel Laureates for that work,[1] but this is not a reason to suggest that peer review failed because it didn't recognize ground-breaking work. Good work will out, and will eventually be published, perhaps with modified conclusions and interpretations.

Some general and ethical points need to be borne in mind in editorial decision-making. The World Association of Medical Editors (WAME; pronounced 'whammy') has formulated some excellent policy statements on this. Although they are primarily aimed at editors of biomedical journals, they are applicable to the editors of all journals. Some are quoted below (with permission from WAME). They (or updated versions) and others can be found on the WAME website (www.wame.org).

1. Editors-in-chief must have full editorial independence
(see Golden Rule 7)

> Editors-in-chief should have full authority over the editorial content of the journal, generally referred to as 'editorial independence.' Owners should not interfere in the evaluation, selection, or editing of individual articles, either directly or by creating an environment in which editorial decisions are strongly influenced.

2. Editorial decisions must be based on the merits of the work submitted and its suitability for the journal, and not dictated by commercial reasons
(see Golden Rule 8)

> Editorial decisions should be based mainly on the validity of the work and its importance to readers, not the commercial success of the journal. Editors should be free to express critical but responsible views about all aspects of medicine

Golden Rule 7

Editors-in-chief must have full editorial independence.

Golden Rule 8

Editorial decisions must be based on the merits of the work submitted and its suitability for the journal; they should not be dictated by commercial reasons, be influenced by the origins of a manuscript, or be determined by the policies of outside agencies.

without fear of retribution, even if these views might conflict with the commercial goals of the publisher. To maintain this position, editors should seek input from a broad array of advisors, such as reviewers, editorial staff, an editorial board, and readers.

3. Editorial decisions should not be influenced by the origins of a manuscript or be determined by the policies of outside agencies (see Golden Rule 8)

Editorial decisions should not be affected by the origins of the manuscript, including the nationality, ethnicity, political beliefs, race, or religion of the authors. Decisions to edit and publish should not be determined by the policies of governments or other agencies outside of the journal itself.

Divided opinions from reviewers

There will be some manuscripts that receive what appear to be opposing reviews. In some cases they may be so polarized that whoever is receiving them might wonder if the reviewers reviewed the same manuscript. That they did should be checked before the reviews are transferred, or their arrival notified, to whoever will be making the editorial decision, just in case the reviewers either didn't review the same manuscript because of an error on someone's part (see Chapter 4, page 66) or they didn't submit the correct review or report and this wasn't spotted when the review was submitted (see Chapter 4, page 75). It is not unknown for reviewers to upload the wrong file to the reviewing form. The detailed comments should provide enough information for this to be checked, but if there is any doubt, the reviewer should be contacted for clarification.

Once it is clear that the opposing reviews do refer to the same manuscript, it is not, for the following reasons, good practice automatically to send the manuscript out for review to another reviewer on the original reviewer list:

1   If the editor's own area of expertise is very close to that of the work in the manuscript, they may be able to make an editorial decision based on the reports that are in. From their knowledge of the reviewers and the work, they will know the weighting that should be attached to the two reports. One of the reviewers may be a real expert whose opinion counts for more. There may be some conflict of interest in one of the reviews. In some cases, the two reviewers may have been called upon to review different aspects of the work, so divided opinions may not be unexpected. The editor should be able to assess the degree of severity or unimportance of the criticisms. Reports of flaws or technical criticisms need to be taken more seriously than subjective opinions on suitability for the journal. As always, the actual recommendation isn't the most important thing to consider – it's the detailed comments that need to be assessed and evaluated by the editor, as was done by the editor who wrote the following to an author:

> Although the recommendations differ, all three reviewers have essentially identified very similar problems and made nearly identical suggestions for the revision of your manuscript.

2   One of the reviewers may, for a number of reasons, be totally off-beam in their assessment and criticisms or enthusiasm – the editor may be able to spot this.

3   One of the reviewers may have provided a 'non-review', i.e. one that is inadequate or superficial – perhaps a case of the reviewer not having read the manuscript properly. This should be readily apparent to the editor.

4   Before doing anything else, the editor may want the authors to be contacted for their responses to the negative reviewer's comments, and perhaps be asked what they would do to address the criticisms made by this reviewer. There might be just a few issues that are crucial. It may be that poor presentation or omission have made the manuscript ambiguous, and confused the reviewer. The author may be able easily to provide clarification. Alternatively, the issues may not be resolvable.

So, before any further action is taken, the editor (or whoever is handling the review) should be contacted, as their input and expertise at this stage are vital to ensuring a high standard of peer review. If the editor is away and not dealing with editorial work during that time, editorial staff will need to try to assess the situation, or perhaps call on the help of another editor. With some editors it may be possible to plan with them before they leave what they would like done in cases of polarized reports. They will likely have different instructions for different manuscripts. They may have been expecting some manuscripts to have a rough ride through review, and they may already have thought about plans of action for different eventualities.

Certain editors may, in the interest of speed, ask that manuscripts with divided opinions routinely be sent to a third reviewer from their original list, i.e. without referring back to them. This calls for some diplomatic feedback, particularly when dealing with relatively inexperienced editors, who may not be aware of the subtleties involved. Such a global instruction can lead to a prolonged and inadequate review, and therefore disgruntled authors.

Beware!

Do not automatically send out for further review a manuscript with divided opinions from the reviewers.

If it is decided that further review is required, the following information will be needed:

1   *Who, and how many people, should be approached to review?* The original list of reviewers may no longer be appropriate. The reviewers' reports already received may highlight specific deficiencies or concerns, and it may be important that other people with specific areas of expertise are approached at this stage.

2 *Should the new reviewers be told that the manuscript has already been reviewed and received mixed opinions?* If they are to be told, should they have sight of these reports (anonymous unless open review is used) or not?

3 *Should the new reviewers be asked to concentrate on certain aspects or be asked specific questions?* The following is an example of excellent guidance received by editorial office staff from an editor regarding a third review after the two original reviewers submitted differing opinions but one reported a potentially serious flaw:

> I have some serious questions about the analysis employed in this paper. But I am not expert in this kind of analysis. I would therefore like to get one of three people to review this: X, Y or Z. What I would like to do is have them see the comments of the two reviewers and specifically ask whether the data analyses are valid. I am particularly concerned about Reviewer 1's comments about the 0.05 cut-off value leading to misidentification of as many as 400 genes. What he/she says makes sense to me and has major implications.

Editors should be encouraged to provide the information listed above in points 1–3 when they request a further reviewer – it will save everyone time and possible aggravation.

If a manuscript is to go out for further review, it's a good idea to let the authors know what's happening. It will stop them worrying and possibly complaining, to others as well as to you, about how long the review process is taking. Be prepared that they may ask to see the reports already received. Many journals do not disclose reports until the review process has been completed, but this is policy on which each journal needs to decide. Be aware, though, that premature release of reports may lead to authors submitting lengthy rebuttals, and even modified versions of the manuscript, before the review process has been completed.

Beware!

Premature release of reports may lead to authors submitting rebuttals or modified versions of the manuscript before the review process has been completed.

Some journals may decide to set up a group of reviewers who adjudicate in cases of divided opinions, especially if the general standard for acceptance is being ramped up. However, whether this is feasible for particular manuscripts will depend on whether acceptance hinges on technical issues or on novelty and general suitability for the journal. In the former case, more external specialist reviewers may need to be consulted to resolve the issue. In the latter case, there needs to have been editorial discussion to determine the scope and standard of paper the journal requires for acceptance and this needs to have been translated into clear guidelines which the special group of reviewers can use in coming to their recommendations (see Chapter 3, page 26).

The decisions that can be made

Most journals have on their reviewing form categories of possible recommendations the reviewers can select. The editors will also usually make recommendations from the same categories, but they will in most cases be able to incorporate subtleties and add provisos. There are a number of possible decisions. It is important that it is made clear to every author exactly what the decision is on their manuscript, the reasons for it, and, if appropriate, what conditions need to be met for the journal to consider the manuscript further either then or at some time in the future. What are the possible decisions after review (see Table 5.2)?

1. Acceptable as stands

Acceptance as stands does occur after a first round of review, but it is very rare for an initial submission as most manuscripts, however good, will receive some suggestions for improvement from reviewers. Resubmitted manuscripts may, however, warrant this decision as they will already have been extensively revised in response to feedback from reviewers and editors on the original submission.

2. Acceptable with minor revision

Manuscripts that are acceptable with minor revision will usually require changes such as minor rewriting, clarification of points, correction of citations, addition or deletion of details, or improvement of figures. They will generally be changes the authors can make without doing extra experimental work or just that which the editor knows will take a relatively short time. Editors should make clear whether all the revisions suggested by the reviewers need to be done or if some can be left out. They should also add any revisions they themselves would like made. To avoid any misunderstanding, it should be stated that acceptance is provisional, and conditional on satisfactory revision. Final acceptance should not be given until satisfactory revisions have been made and the final version submitted. It is wise to put a time limit on submission of revised manuscripts in this category. If there is no time limit, journals may find themselves with manuscripts submitted after lengthy delays, during which time similar manuscripts have been published elsewhere or a field has moved on and the results, or technology used, are no longer as publishable as they were at the time of decision. One to two months is a common time (for many fields, as extensive experimental work, which would take very different times to complete in different disciplines, does not have to be allowed for in this category), long enough to give authors a reasonable chance to deal with all the revisions and to do them well, but short enough for the opinions and points made during review still to be relevant and valid.

Beware!

Make sure authors understand that acceptance for 'acceptable with revision' decisions is not automatic, but conditional on the revisions being satisfactory and any concerns being sufficiently addressed.

Table 5.2 The decisions that can be made on a manuscript after review.

Decision	Features	Editorial action
Acceptable as stands	Occurs, but rare for a first submission.	Make sure all items required for manuscript to move to preparation for publication have been received.
		Make sure standard of language is acceptable.
Acceptable with minor revision	Manuscript requires minor changes such as some rewriting, clarifications, correction of citations, addition or deletion of details, improvement of figures. Manuscript does not need extensive extra experimental work.	State acceptance provisional and dependent on satisfactory revision.
		Make clear all the revision requirements, including whether there is anything the reviewers have requested or suggested that does not need to be done.
		Request point-by-point responses to the reviewers' and editor's comments, and reasons for any revisions not made. Also details of any additional changes made.
		List enclosures and whose, if any, signatures are required.
		Set time limit for submission of revised manuscript.
		Provide full submission instructions for revised manuscript and provide all necessary forms, etc., or information on where these can be found.
Acceptable with major revision	Manuscript requires extensive reworking or additional experimental work.	Take care and do not use this decision if extensive additional experimental work is required.
		If decision is used, editorial action as for 'Acceptable with minor revision'.
Rejected with resubmission invited	Manuscript has significant problems or requires extensive additional experimental work.	Give the conditions under which a resubmission will/will not be considered.
		Make clear exactly what revisions are/are not required and whether there is any time limit for submission.
		Outline procedures for handling/review of any resubmission.
		Request point-by-point responses to the reviewers' and editor's comments, and reasons for any revisions not made. Also details of any additional changes made.
		Relay appropriate level of enthusiasm or otherwise.
		Provide full submission instructions and list enclosures required.
Rejected with no encouragement to resubmit	Manuscript not acceptable for publication in the journal and not likely to become so in the future.	Make clear why the manuscript is being rejected.
		If work is sound but just not of great enough interest or significance to be accepted, use sensitivity and diplomacy to minimize disappointment and help maintain good relations for future submissions.
	May have serious problems, be flawed, of too low interest or significance, or too specialized.	Advise on alternative journals if appropriate.
	Submission may have involved misconduct or misbehaviour.	If there have been any allegations or suspicions of misconduct, do not ignore; look into at journal level (see Chapter 9).

Box 5.1 Items authors might be required to include on submission of their revised manuscript

■ The revised manuscript, one copy with all changes highlighted in some way, and one 'clean' copy. Details should be given on how these should be submitted – online, by email, as digital copies on disk or CD (both online and CD may be required) – and whether hard copy printouts are needed. If separate files need to be submitted for the various parts of a manuscript, instructions should be provided on this. File format and maximum size should be indicated.

■ A set of reproduction-quality original figures. If digital files are required, details of file format needed and maximum size permitted should be given. Assistance with figure preparation for publication is helpful to authors and increases the likelihood of satisfactory and usable figures being sent. Even with online submission systems, hard copy figures may be required, as digital files are not always adequate or hard copy versions may be needed for correct colour matching.

■ A full response to the reviewers' and editor's comments, detailing point-by-point the changes made and giving reasons for those not made. Details of any additional changes made.

■ Completed and signed forms that have not yet been received, for example for copyright assignment, exclusive licence agreement, or agreement to pay colour-work charges. Information should be provided on where the forms can be found (for example, on the journal's or publisher's website) if blank copies are not sent to the author.

■ Copyright clearance for material used from third parties.

■ Confirmation of any outstanding database or material repository depositions, with appropriate accession and reference numbers.

■ Possible cover images. Instructions on how they should be prepared and sent should be provided.

When authors are notified of the decision, they should be advised on all practical matters such as the time allowed for revision, how and where to submit the revised manuscript, what ancillary materials are required, and whether any hard copy materials need to be sent (see Box 5.1). Organization at this stage will pay off and avoid delays both in the final decision being made and in preparing the manuscript for publication.

3. Acceptable with major revision
This can be a tricky category and manuscripts falling into it need to be handled carefully and thoughtfully. There will frequently be an 'accept with major revision' category on reviewing forms, but journals may decide not to express decisions in this

way in their outcome letters to authors. If the major revision involves just a major restructuring and/or rewriting without additional experimental work, then a manuscript can be provisionally accepted pending required revisions, as it is clear it will be possible for the authors to deal with them in a defined timeframe. If the major revision involves extra experimental work this can lead to problems, as the results may take a long time to obtain, for example either because of the methodology used or because material needs to be collected or grown. Such manuscripts may also need to be reviewed again in a full review process. It is better to reject such manuscripts and invite a resubmission (see point 4 below). Alternatively, if editors suspect the authors may already have the extra data, or be able to obtain them quickly, they could offer a provisional-accept decision (as in point 2 above) but stipulate clearly that if it does not prove possible to submit the revision within a certain time, the decision will change to a 'reject with resubmission invited'. Clarity is of the essence and there should be no confusion about what the decision is or what the requirements are.

4. Rejected with resubmission invited

There will be manuscripts that editors cannot accept at the present time because of problems that have been identified in review, but which they would like to see again at a later date, as a resubmission, if certain issues can be addressed. Their enthusiasm for seeing a resubmission will vary and this needs to be conveyed to the authors, firstly out of fairness to them, so they can decide whether they would be better off submitting their manuscript elsewhere, and secondly so that the journal does not lose manuscripts that have the potential to become very good papers. Sensitive use of language, for example using the word 'decline' instead of 'reject', can also help convey to authors a hopeful future outcome rather than a sharp and final negative decision.

Beware!

Take care to use appropriate language and terms when conveying negative decisions.

Most manuscripts in this category will have received reviews either that are lengthy and substantial, and require the authors to address many points, or that require extra experimental work, perhaps extensive, to be done. It is crucial that editors stipulate exactly what authors need to do and what they can ignore. Some reviewers' requests may be totally unreasonable and involve many years of extra work. It is for the editors to arbitrate and decide on this (see Golden Rule 5).

Manuscripts in this category may also have received divided opinions from the reviewers, but the editor has decided to follow the negative reviewer (see this chapter, page 91). In these cases, it is particularly important that authors are given the reasons why the decision has gone the way it has. If they aren't, not only might they feel they have been treated unfairly, they may also send in a rebuttal claiming that it is the positive reviewer who is correct and asking for the decision to be reconsidered.

The following is an example of a decision letter to an author in which the editor provided full reasons for the rejection decision:

> You will see that the reviewers came up with divergent recommendations. After carefully weighing their comments, and thoroughly evaluating the manuscript, I am sorry to inform you that I agree with the reviewer who recommended that the paper be rejected. This was not an easy decision, because there is much in the paper that I think is useful and informative. But in the end, I feel that the interesting observations represent an incomplete story. I realize that in this field it is often very difficult to get complete data, and allowances for such things should be made. But after weighing the totality of the material you and your colleagues present in this manuscript, I feel a negative decision at this point is justified, as I explain below.

The editor then went on to list in detail the technical and other points that had led to the final decision. The authors readily accepted the decision and were grateful for the editor's comments and insight. The manuscript was resubmitted greatly improved, and accepted for publication after another round of review. The peer-review process worked as it should, everyone was happy, and good, sound work was published.

Editors, as well as detailing exactly what revisions will be required for a manuscript to be considered again, should provide authors with some idea of what the procedures will be for handling a resubmission of their manuscript – for example, whether it will be sent to the same or new reviewers, or whether perhaps a combination of the two will be used. This may influence author decisions on how to move forward. If new reviewers are to be used, they may decide to submit elsewhere because of concern that new points may be raised. If the editor makes it clear that, if possible, the same reviewers will be used, authors know that they will have to satisfy those reviewers with the changes they have made, or with reasons why they have not made some. If the authors know they won't be able to address the issues raised by the reviewers, then, again, they may decide to submit elsewhere. If the editor is a real expert in the subject area of the manuscript, he or she may indicate that they will assess the manuscript themselves as long as the authors submit a very clear cover letter detailing the changes they have and have not made. This will sharply focus the authors' minds, and they will usually do their utmost to satisfy the editor's and reviewers' requests.

Each journal needs to decide whether it wants to set a time limit for submission of resubmissions. Frequently, extensive experimental work may be required, and it may not be easy to gauge how long this will take. Manuscripts can be submitted a number of years later and still be relevant. They may even be vastly improved and have been extended to present a more complete story. On the other hand, a field may have moved on and the work be out of date by the time a manuscript is ready to resubmit. In cases where a manuscript might be made acceptable after major revision, an editor may want to enter into a dialogue with the authors before finaliz-ing a decision, to find out whether they could do certain things or how they would address certain of the reviewers' comments. This can be very productive and help formulate the requirements for possible eventual acceptance of a piece of work.

5. Rejected with no encouragement to resubmit
There will be manuscripts that need to be rejected and which the editor, for one reason or another, does not want to see resubmitted. Reasons could include: the work is so flawed that there is no prospect of the author being able to salvage it; the results fall below the interest and/or significance thresholds for acceptance and cannot be raised above the thresholds; the work is too specialized; serious problems (such as fraud or plagiarism) became apparent during the review process. If any potential misconduct did come to light in the review process, then this should have been investigated (see Chapter 9, page 184). If the work is sound but just not appro-priate for that journal, recommendations can be given on the journals for which the manuscript may be suitable.

Checks to be made before communicating decisions to authors

Before any decision is finalized and communicated to the author, a series of checks need to be run to make sure that everything has been taken into account and that the decision is consistent with journal policy and with other decisions being made across the journal. Whoever has responsibility for communicating decisions to authors will need to make various checks.

1. The correct number of reviews are in and have all been considered
by the editor
Most journals routinely send manuscripts to the same number of reviewers, most commonly two (see Chapter 4, page 52). Occasionally, however, there may be either more or fewer. The editor may miss this, and submit their recommendation or deci-sion prematurely if more reviewers than usual have been used, or not submit it but be waiting for a non-existent review if fewer than usual reviewers have been used. Both these situations introduce delays, and procedures must be put in place to avoid them. If editors are dealing with receipt of reviews themselves, they need to be alert to the potential problems. If a central office is dealing with reviews on an online system, then editorial staff can notify editors with a brief message when all reviews are in for a manuscript and ready for their attention. They can also at the same time update editors about special circumstances or remind them of important relevant information. If hard copy reviews are still being used in a paper-based system, and being submitted to a central office, they can routinely be transmitted to the editor

when they are all in, or as they come in but with a note attached to the last to indicate it is the final one. If reviews are being submitted direct to editors, their return
will need to be monitored by them and decisions submitted at the appropriate time.
If an editor submits a recommendation or decision prematurely on an online system,
it may in some cases prevent the remaining reviewer(s) from being able to submit
their review(s). If this cannot be dealt with by the editorial office, the online service
provider needs to be contacted as soon as possible to make the technical adjustments to the database to allow the remaining reviewer(s) to submit their review(s).
This not only introduces delays into the review process, it may also frustrate reviewers and there is a danger that there may be further delays waiting for them
to attempt to submit their reviews again. If any reviews are long overdue, editors may
decide not to wait, but to make and submit their decision on the basis of the reviews
that are in. They should let the editorial office know they are doing this, either with a
message or by adding a note to the manuscript notes screen if an online system is
used, so that staff are aware the action was intentional and made in the knowledge
that not all the reviews were in. Similarly, editorial staff should notify editors when
reviews are overdue and proving impossible to obtain (see Chapter 4, page 71).

Beware!

Make sure editorial recommendations or decisions haven't been submitted
prematurely.

2. The editor has submitted the right letter for the manuscript
Editors may occasionally download the wrong document when submitting decisions,
especially if they are dealing with a number of manuscripts at the same time, or the
same authors have more than one manuscript under consideration at the same time.
They may also, by mistake, be working on the wrong manuscript screen and this can
lead to confusion unless this is spotted and corrected or an alert note added to the
manuscript notes. If editors are on the wrong manuscript screen and communicating
with authors and reviewers themselves, they may also end up emailing the wrong
person or people. Keeping the '3N' Rule (see Chapter 2, page 22) in mind can help
prevent this. A quick check of manuscript number, title, authors, and reviewers'
reports against the decision letter content should be done to see if they appear to be
consistent. If completely different nomenclature or terms appear, the editor needs
to be contacted. Letters should also be checked for completeness to ensure that
parts haven't been left off when an editor has been copying and pasting from an
electronic file.

Beware!

Editors can sometimes download the wrong document for a decision letter.

Beware!

Watch out for truncated or incomplete letters from editors resulting from errors in copying and pasting.

3. Inappropriate material is removed or amended
Occasionally, an editor may write notes within the decision letter for the authors that are intended just for the person communicating the decision – for example, requesting that wording be changed to convey a particular tone, or that a fact needs to be checked if something is to be included. These need to be acted on and removed when preparing the letter. Ideally, it's best that editors don't add such comments to their letters, but rather make them elsewhere to avoid any chance of them being left and so transmitted to the author. Editors may also inadvertently mention the reviewers' names within their letters, and unless open peer review is being used by the journal (see Chapter 4, page 42), these should be removed and replaced with the identifying number or letter of the reviewer, i.e. 1, 2, or A, B, etc. Words or phrases that might cause confusion, or offence because of cultural considerations, should also be changed.

Beware!

Watch out for reviewers' names and notes and instructions for editorial staff in editors' decision letters for authors.

4. The recommended decision is appropriate and consistent with journal policy and decisions on other manuscripts
If the decision an editor is recommending doesn't seem to correspond to what they've written in their letter, he or she should be contacted for clarification. They may just have selected the wrong decision in error when ticking a checklist. Editorial staff should be able to spot this in most cases, as it doesn't usually involve specialist knowledge of the subject of the manuscript. Ideally, journals should have in place a mechanism to check consistency of decisions, and if the decision on any manuscript appears inconsistent with the decisions being made on other manuscripts or by other editors, the editor in question should be notified and given feedback on why this is so. Access to the other manuscripts and reports may be necessary and/or discussion with the editor(s) dealing with those manuscripts to ensure that decisions are both fair and consistent. Inconsistency in decisions being made on manuscripts based on scientific criteria or quality judgements is not, however, something general editorial office staff are able, or should be expected, to assess. This requires specialist knowledge and experience, and journals need to decide what mechanisms to put in place for this. Journals should do all they can to make authors feel confident that decisions have been made as fairly as possible. All decisions can go through one

person, for example the editor-in-chief or one of the editors (see Table 5.1). Individual editors should also be given guidance on decision making, particularly when they first join a journal, provided with updates as editorial policy changes, and advised to flag up any decisions that they find problematical or are unsure about so that other editors can be brought in to give opinions. It is easy to say that journals should monitor consistency in decision making, but in reality consistency is difficult to achieve, especially as for many journals there will be a large number of manuscripts that fall within the 'grey' area between definite accepts and definite rejects. Here there will be many manuscripts of similar quality, and acceptance will hinge on subjective assessments of novelty, usefulness, or other criteria defined by the journal. This is where the quality and calibre of the editors comes to the fore, and journals which have selected editors wisely can rely on their knowledge and experience to make close judgement calls.

Communicating the decision to the authors

Once the editorial decision has been finalized it needs to be communicated to the authors. This should be done in a constructive way and all the necessary enclosures and instructions included if a revision is being invited (see Box 5.1). The decision and supporting documents should be sent to the appropriate author, the corresponding author, or to someone they have nominated, for example if they are away or not themselves able to deal with editorial correspondence. Occasionally, a corresponding author may, on submission, ask that all authors be advised of all communications; some journals may go along with this, but it can become very time-consuming and in general it is better to communicate with just one author and leave it to them to be responsible for passing on information and the outcome to their co-authors. With online systems it is actually very easy to set up the systems to relay correspondence to all the authors. However, it is also very easy for all these authors to press the 'reply' button and a journal may end up receiving messages from various co-authors without them having co-ordinated their response in any way. Messy, as well as time-consuming, situations can result. Occasionally, in joint communications, authors may think they are communicating with just their co-authors, press 'reply all' and include the journal in correspondence they would probably prefer it didn't see! Some revealing, and sometimes embarrassing (for the authors), messages can be received.

Beware!

Take care with requests from authors that communications for their manuscript be sent to all the co-authors – deal with just one author, the 'corresponding author'.

Online systems automatically log decisions and the date sent, and forward the appropriate documents to the authors. If stand-alone or home-grown tracking systems are being used, the decision and date of transmittal to author need to be noted, along with any special instructions for further action.

If any of the reviewers have included a hard copy of the manuscript for the authors marked up with corrections and comments, it should be checked carefully to make sure there are no accidental enclosures (for example, sticky notes with messages for the reviewer or others made by the reviewer, shopping lists, pages from lecture notes or grant proposals – all of which I've come across). If a reviewer has submitted a marked-up portable document format (PDF) file or other electronic version, all identifiers to the creator must be removed if that hasn't been done by the reviewer (see Chapter 4, page 74).

Rebuttals and appeals from authors

All journals, however well run and fair their peer-review process, will receive rebuttals and appeals from authors. There may be a few authors who always seem to appeal against negative decisions as a matter of course. Some may do this so frequently, and usually without good grounds, that when they do have good cause to appeal the impact of their appeal is diminished. On the other hand, there are authors who are generally very reasonable and accept negative decisions that they consider to be fairly reached even though they may, understandably, be very disappointed. When such authors register an appeal, it does need to be taken seriously. Journals should have a formal system for appeals. Some, like the journals of the American Physical Society (see www.aps.org), give details of their procedures in their editorial policy guidelines. Others, however, feel that to advertise such a system is to invite appeals. An appeal should really be viewed as a dialogue and a continuation of the peer-review process. The claims and points made by the authors should be considered and evaluated. Editors may feel they need to contact the reviewers again for further advice. How the reviewers respond will be influenced by how the author goes about presenting a rebuttal. Editors and editorial offices may be called upon by authors for advice on how they should go about this. In the interests of ensuring a smooth and fair process and maintaining good relationships between the authors and reviewers, they should know what advice to give. Authors should not, even though they may feel like it, insult the reviewers by, for example, claiming that they (i.e. the reviewers) know nothing about the subject or have totally missed the point because of some deficiency on their part. Starting a letter with a polite and neutral sentence such as, 'We would like to thank the reviewers for their constructive criticisms and comments, which will undoubtedly help us to improve our manuscript. However, . . .' is much more likely to ensure the reviewers (and editor) are receptive to what the authors then go on to say and to considering the points they make. The authors should present their case calmly and objectively. They should deal with all contentious points one by one, and provide evidence to back up their claims, or appropriate clarification if it is clear the reviewers have missed or misunderstood something, admitting perhaps that this has probably occurred because of problems associated with the presentation of the manuscript in the first place.

Beware!

Don't automatically dismiss appeals from authors – reviewers, and editors, can be wrong.

Editors should assess appeal or rebuttal letters from authors on a case-by-case basis. They should not dismiss them out of hand, saying they don't ever consider appeals and that their decisions are final. The authors may have very valid points to make and the reviewers might have made mistakes or been wrong about certain things. After considering the points, an editor may decide to consult with one or more of the other editors or all or some of the reviewers, to send the manuscript out to a new independent reviewer or reviewers for a further opinion (either with or without the appeal material), to change or qualify the original decision, or to stand by it. If the last is the case, the editor needs to make it clear to the authors why he or she is doing this, and to give full reasons. Authors will more readily accept a negative decision if they feel it has been arrived at fairly, even if they still don't agree with it.

Authors base their appeals on different evidence. Those based on scientific evidence need to be judged on the merits of that. However, a common appeal is to contest a decision on the basis of what the journal has published previously. Some authors may even go back years to find examples to back up their claims. Unless the papers referred to are very recent, this is one of the weakest of arguments, especially for journals that report novel research in fast-moving fields. Things may have moved on rapidly, and what would have been considered novel and worthy of publication a year or so ago will no longer warrant publication in that journal. Editors should not be surprised to receive appeals in fields where research is expensive and time-consuming, where the authors may have invested very large amounts of both resources and time in producing a paper and are desperate to publish it in as high an impact journal as possible. Other journals will still be interested in publishing such work if it is good and sound, and it is for the authors to identify the best and most realistic venue for that piece of research.

Dealing with revisions

The majority of authors whose manuscripts have been provisionally accepted pending minor revision will go on to submit revised manuscripts, as the revisions will have been gauged by the editor to be both feasible and capable of being completed in the time allowed by the journal for revision. But no matter how important it is to an author that their manuscript be published, they may lose track of time or be so involved with other commitments that they end up in danger of missing the revision deadline. It is therefore a good idea, and usually much appreciated by authors, for journals to send out a brief email reminder a short while before the deadline, for example a short message such as that given below:

> Dear [Author],
>
> You are currently revising the above manuscript. To help us monitor manuscript status, I would be grateful if you could let me know as soon as possible approximately when you will be submitting the revision to us. Can I please remind you that revised manuscripts must be received by us within 2 months of the date you were notified of the decision. You were invited to submit a revised version of your manuscript on [date].
>
> Thank you in advance for your help, and I look forward to hearing from you.

This serves a number of purposes. Firstly, it reminds the author that the revision deadline is approaching; secondly, it tells them what the deadline date is – some will inevitably have forgotten it; and thirdly, it invites a response from the author, and this will usually come in quickly. Most authors will just confirm that they are revising and give an idea of when the journal can expect to receive the revised manuscript. This is very useful because it helps in estimating potential copy for future issues. It also gives the authors the opportunity to bring to light any problems. Some may have found that they were not able to do the revisions and so decided to submit to another journal; some may feel they need extra time to do the revisions and ask for an extension. Requests for extensions need to be evaluated on a case-by-case basis and not just given whatever the reason. Authors will soon learn that a journal is over lenient and stop viewing the deadline as a reality. They may feel no pressure or obligation to submit their revision on time, which can have negative effects and occasionally cause chaos, especially if a journal is aiming to publish certain papers together. But if there are genuine reasons why there is likely to be a delay – for example if an author is in the process of moving labs, has been ill or suffered a personal tragedy – then journals should look favourably at granting an extension. A new date should be agreed rather than a vague 'yes, fine' given in response to the request, or things may go awry and a revision not be returned for a long time, during which time the work may have been superseded and have lost some of its novelty or value. If things look as if they may drag on, a journal needs to reassess the situation and perhaps offer new conditions for revision and acceptance, for example being prepared to consider a revision if it can be submitted before a certain time but that after that the manuscript will need to be officially declined and treated as a resubmission if the authors still want to publish it in that journal. If there are any changes in the conditions that need to be fulfilled if the status changes to a resubmission, these need to be made very clear to the authors to avoid the risk of confusion and perhaps accusations of unfairness later on.

> ## Beware!
> Avoid being too lenient in extending revision deadlines – authors may stop taking them seriously.

Care should always be taken when using the term 'revised manuscript' to avoid ambiguity in cases where not just a straightforward revision is being invited. Authors may focus on the word 'revised' and mistake an editor's intended meaning when, for example, a resubmission (of a new manuscript after appropriate revision) is actually being invited. It should be made clear if the current manuscript is being provisionally accepted pending satisfactory revision, or if it is being rejected at the present time but a new, resubmitted, manuscript is being invited.

> ## Beware!
> Take care when using the term 'revised manuscript' with authors – they may misunderstand it.

If a journal has given clear instructions in the editorial decision correspondence and the author has followed those, it should receive all the appropriate materials and enclosures with the revised manuscript. Revised manuscripts accepted subject to only minor revision can usually be assessed by the editor without the need for the reviewers to be consulted, as the revisions needed are generally straightforward. If authors have not highlighted changes in the revised manuscript they should be asked to provide such a copy, as it makes assessment much easier and more accurate. A complete and concise point-by-point response to the reviewers' and editor's comments will greatly aid assessment, and if it hasn't been sent the authors should be contacted for one. Editors should not issue a blanket instruction that all revised manuscripts should be sent to the original reviewers for assessment. The decision as to whether external review is needed at this stage should be made on a case-by-case basis, and reached only after looking at the revised manuscript and considering the authors' responses. Editors who routinely send all revised manuscripts out for checking, even if there may be only the most minor of changes or if the authors' responses require the editors' attention, risk annoying the reviewers, who may feel they are doing the editor's work for them. However, if a manuscript is not in an editor's direct area of expertise, or if the editor is concerned that some of the reviewers' points may not have been dealt with adequately, then they should send the revised version for assessment to one or more of the original reviewers, either for a full review or just to look at certain points. Simple textual changes can be checked by editorial office staff if requested by the editor. However, if the staff are not comfortable doing this, are concerned about any of the changes, or if some changes have not been made or additional ones to those requested have appeared, they should refer the manuscript to the editor.

If the authors have fulfilled all the requirements requested by the editor, and within the stipulated time, then the manuscript should be accepted. If they have not made all the necessary revisions, or have not been able to justify adequately why they haven't made them, the editor may decide either to reject the manuscript or to ask for a further revision. The latter will usually be relatively minor, but may include some important points; it may also involve authors incorporating some of the things they have included in their letter of response into the manuscript itself or adding them to supplemental material for online publication (see Chapter 3, page 31). Communicating the final decision should follow the same guidelines as described previously (see this chapter, page 102).

Dealing with resubmissions

There are basically three types of resubmissions:

1   Those from authors who were originally given a 'provisionally accepted with minor revision' decision but were unable to make the revision deadline.
2   Those being submitted in response to encouragement to resubmit following a rejection decision.
3   Those being submitted by authors whose original submissions were rejected and where resubmission was neither invited nor encouraged.

For the first two cases, the procedure is fairly straightforward as the authors should have been given clear guidance on what revisions were needed and the conditions under which their manuscript would be considered again and could be accepted. So there is an agreement framework in place and the editor should abide by this. In the last case, however, an editor is not under any constraints and can view the manuscript as a completely new submission without any obligations.

How should assessment of these three types of resubmissions be dealt with?

1. Revised manuscripts that have missed the revision deadline
The editor may well be able to assess the resubmission, especially if he or she is a specialist and/or it is not long since the original submission was under consideration. If the editor is not particularly close to the subject area, or it is some time since the original submission, he or she may want to send the manuscript out to reviewers, usually to the same reviewers as were used previously, or perhaps to just one of them. Who is chosen will depend on the revisions that were needed and on what is in the authors' letter of response.

2. Invited resubmissions
The editor may be able to assess the resubmission if the work falls within his or her expertise, particularly if the authors have clearly described the changes they have

or have not made and have submitted a comprehensive letter in response to the reviewers' comments. If the editor decides the manuscript needs to be seen by the reviewer(s), who it is sent to will depend on what the editor told the authors would be done when the original decision was sent to them, on the revisions made, and on the responses received. Generally, it is a good idea for resubmissions to go back to the original reviewers – these individuals are familiar with the work and recommended the revisions required and highlighted the problems. The considerations put forward when discussing whether or not to get a reviewer's agreement to review (Chapter 4, page 53) still hold, but there is the complication that if the original reviewers refuse to assess the manuscript new reviewers will need to be contacted and they may raise new criticisms of the work and request new revisions. Journals may therefore in such cases decide to send the manuscript to the reviewers without first checking that they are willing to review it; it is up to each journal to decide what works best for them. Return without invitation is warranted if the journal has questions on this in its reviewing forms and the reviewer previously ticked that they would be agreeable to assessing a further version. If reviewers are first contacted, it should be mentioned that the manuscript is a resubmission of one they previously assessed. This will generally, unless they had to struggle to review the original version, make reviewers more willing to take a look at the new manuscript as they will already be familiar with it. Some reviewers will actually be keen to see the revised resubmitted version to make sure the authors have done what they were supposed to before possible publication. Journals should be aware of 'reviewer fatigue' in cases where reviewers have seen a number of versions of the same manuscript and indicate they do not want to see another version. This may be because they are just fed up with seeing it and perhaps feel they can no longer carry out a fair assessment, or they may be annoyed because the authors seem to spend more time arguing and justifying why they haven't made the revisions than working to incorporate them. If reviewers don't wish to see any further versions of a manuscript, their wishes should be respected.

Reviewers assessing invited resubmissions should ideally have certain materials made available to them to ensure a fair and thorough review:

1   The resubmitted manuscript.

2   The original manuscript.

3   The reviewers' comments on the original submission.

4   The decision letter sent to the author stipulating the revisions needed and the conditions to be met for acceptance.

5   Details of any further relevant correspondence if the revisions and conditions for acceptance were modified post-decision.

6   The authors' letter of response to the reviewers' and editor's comments.

7   All the journal's usual instructions, reviewing forms and checklists (see Chapter 4, page 60), together with any specific guidance on the review of the resubmission.

3. Non-invited resubmissions

As no conditions were laid down for a resubmission if none was invited, editors are free to treat non-invited resubmissions in whatever way they consider is most appropriate. If it is clear that all or most of the original problems still exist, they are at liberty to return a resubmitted manuscript immediately without any further review – sometimes, understandably, with expressions of annoyance that the author has ignored the previous review and is wasting the editor's and journal's time. Alternatively, the field of research may have moved on considerably in the time since the initial submission and a manuscript may no longer be appropriate for the journal. The editor may in some cases, however, if it is clear that revisions have been made and criticisms addressed, decide to consider a manuscript further and send it out for review. Some or all of the previous reviewers may be used, or all new ones, or a combination of the two. The editor needs to decide whether or not any new reviewers should be made aware that an earlier version of the manuscript has already been considered. It may be best to treat it as a completely new submission. If a resubmission wasn't invited but the reviewers are made aware that it has been considered before, care should be taken not to give the reviewers any impression that it has been invited as they may, perhaps subconsciously, feel more positively towards it, thinking that if the editor encouraged resubmission some of the reviewers' negative and possibly more subjective comments can be ignored. It is important that the reviewers submit a review based on a complete and unbiased assessment of the current submission.

Beware!

Make sure reviewers of non-invited resubmissions are aware that the resubmission wasn't invited.

Problems with resubmissions

Resubmissions can sometimes be problematical. An author may resubmit a manuscript and not indicate that it is a resubmission, even if specifically asked about this during the submission process (and it is a good idea to request this information). This may not be spotted, especially if a journal is dealing with a large number of submissions. The title may even have changed, so a title check, for example using the search facility of online systems, won't identify such manuscripts as resubmissions. The situation may be discovered only by chance, for example if the new manuscript is dealt with by the same editor, the same reviewers are selected and approached, or another editor spots it as a result of reports that have been circulated on submissions and rejected manuscripts. If there is any suspicion that a manuscript is a resubmission, searches for the authors' names on the journal's online system or manuscript database will pull up the manuscripts they've previously submitted to the journal.

Normally, the same editor should handle a resubmission because it is they who have set the conditions for resubmission and know the manuscript and its history. Occasionally an editor may already have asked that any resubmission be handled by another editor. They may have been worn down by rebuttals and protests from the authors and no longer feel able to deal with any resulting manuscript fairly and objectively ('editor fatigue'). The new editor should always be fully briefed on the submission history to date and provided with all the relevant documents. In other cases, authors will disclose that their manuscript is a resubmission but request that it be handled by a different editor. The previous editor may not be aware of such a request. Such manuscripts need to be dealt with on a case-by-case basis and the editor-in-chief or managing editor needs to decide how they should be handled; a considerable amount of diplomacy may be needed. It is not good practice to keep such a submission secret from the original editor. They may be upset or annoyed and feel that the journal does not trust them and their decision making. They should therefore be advised of the situation and given an appropriate level of information. Some authors may be making change-of-editor requests for what they feel are legitimate reasons, for example if they feel the editor holds very strong or rigid views that are diametrically opposed to theirs and the two cannot be reconciled, or that the editor does not have adequate expertise in the subject to make the right call. Others may be doing it to increase their chance of a successful outcome, possibly by excluding the real specialist on the editorial board. Authors should never be allowed to play one editor off against another.

Acceptance

A revised or resubmitted manuscript that fulfils the editor's revision criteria and the journal's policy requirements should be accepted. The authors need to be notified in a formal, dated letter of acceptance and the manuscript checked for completeness before moving it on for preparation for publication. Authors need to be contacted for any missing items or if file formatting or size are not as required. They also need to be contacted if it isn't clear that all the journal's policy requirements have been met, for example if final permission for free material availability hasn't yet been obtained from a commercial partner, or database accession numbers are missing. Some journals may want to delay sending formal acceptance until everything has been received from the authors to help ensure they respond and supply the missing items quickly. All manuscripts must meet all requirements before being published, as these were the conditions under which they were reviewed and conditionally accepted. Dates should have been carefully noted throughout the editorial process and the crucial ones (receipt of original manuscript, receipt of revised manuscript, and acceptance) must be forwarded to the production department with the manuscript and these dates should appear on the published paper.

Most journals ask authors to sign an indemnity as part of the exclusive licence to publish or copyright assignment requirements for submission, and authors are responsible for ensuring their manuscripts do not contain plagiarized material or anything that is libellous, defamatory, indecent, obscene, or otherwise unlawful, or that infringes the rights of others. Despite this, editorial staff should check manuscripts before final acceptance to make sure that nothing of this nature is included in the material passed on into the production process for preparation for publication. Most cases will have been (or should have been) picked up during the review process, but this cannot be guaranteed, especially if the editors and reviewers are not native speakers of the language in which the paper is written. Words or text may have been inadvertently used that reflect badly on individuals or their work or papers.

Sometimes an author may contact the editorial office to request changes post-acceptance. Some changes may be trivial and can readily be dealt with by the editorial staff, for example substituting better versions of the figures, making improvements in style or grammar, or correcting simple errors. Some requests, however, may appear to be more significant and possibly to have serious implications. These need to be approved by the editor who handled the manuscript or by another person with authority to make or approve such editorial changes.

The following points describe the sorts of changes that may be requested.

1   *Addition of new data or substitution of existing results with new data.* The editor must be notified of such requests. He or she may be able to deal with them themselves, but in some cases may feel the reviewer(s) need to be consulted before the changes can be approved.

2   *Change in the authorship listing.* Readers are referred to Chapter 8 (page 156), where this is covered.

3   *Requests for a note to be added at proof stage.* Many requests for notes to be added at proof stage need editor approval. The changes may be minimal in extent, but the content may be far from insignificant. The editor may even decide that the request for a note to be added needs to be referred back to the reviewers for comment and recommendation. Those notes that refer to other work that has been published or is in press, either the author's own or that of other authors, need particular care. The dates of submission and publication of the other work need to be checked, as it is not acceptable for authors to put in notes material that should really have been either included in the paper itself or referred to within it. If the related material was publicly available at the time of submission, it should have been referenced and taken into account during review. Most times, the reviewers will have alerted editors to any missing related work and reference to it will have been incorporated at revision stage, but there will be times when it will not have been picked up. Authors may even suggest including in a note in proof a reference to another paper on which there are some authors in common with their own paper – this is not at all acceptable, as

those authors will have (or most definitely should have) been aware of the work being reported in the other paper and should have ensured mention of it in papers being submitted to any other journals. Care must be taken to avoid authors adding notes without attribution that may result from knowledge of papers they have reviewed, as they may be trying to scoop an idea or results. This is very difficult to monitor and usually comes to light only as a result of chance events. If there is any doubt about where new information has been obtained, the author must be contacted for clarification.

Beware!

Take care with requests from authors for changes to their manuscripts after acceptance – seek editorial approval if in any doubt.

Decision making to consistent standards and the problem of availability of space

All journals with a paper version have an annual page budget and this is usually set about half-way through the preceding year. The page budget determines the subscription price, and therefore the income for the journal and so the resources available for the various areas of activity. Online-only journals also have to take into account page budget; even though costs related to paper publishing and distribution are not an issue, there are still costs associated with the selection and handling of editorial content (involving such things as peer review, copyediting and typesetting) and maintenance of the archive. When deciding on whether an increase in pages is required for the coming year, editors need to take into account the increase in manuscript submissions, the quality of submissions, any new policy initiatives that will affect acceptance, and the long-term strategy for the journal.

Page extent should be monitored continuously to ensure that a journal is keeping reasonably close to budget and not building up a backlog of accepted manuscripts that will lead to an increase in publication time. Things can rapidly spiral out of control unless editorial decision-making is regulated. Publication times are important for authors, and sometimes they are critical, especially in fast-moving fields or if the authors know that another group or groups are close to publishing similar work and so scooping them. Journals with long publication times may find that authors will not consider them for their best and most important papers. The advent of article-by-article publication online in advance of print has somewhat alleviated this problem, but has introduced another as some journals have found themselves with steadily increasing times to paper publication as the rate of acceptance produces copy that exceeds the space available in issues.

Decision making should primarily be to consistent standards and not dictated by the availability of space. However, where journals have a large number of manuscripts of similar quality submitted it can be difficult to maintain a common standard. What can be done to help keep within page budget extent but maintain a fair and consistent acceptance/rejection policy?

1   Editorial policy on acceptance and rejection – both scope and quality – needs to be fully worked out, written down, and circulated to all the editors. There should be regular editorial discussion and policy updated as required (see Chapter 3, page 26).

2   Editors should be made aware of what is being accepted and rejected by the other editors as this helps ensure consistency across the editorial team. Reports of accepted and rejected manuscripts should be circulated regularly.

3   Someone must take responsibility for monitoring page extent through the year and alerting those with decision-making powers if any trends appear that need to be addressed – for example if the number of accepted papers is increasing in such a way that the page budget will be reached some way before year end.

4   Radical changes in policy are not usually necessary. Subtle tweaking can avoid the need for this. There will often be a significant number of manuscripts for which the decision could go either way (the 'grey' area mentioned earlier, on page 102). If editors know space is tight they will tend to reject more, and vice versa if the journal has the luxury of being in page credit or has been given extra pages. There is a very fine line between being on budget and being considerably over. For example, if a journal has 12 issues a year, each containing 10 papers that are 10 pages long, that is an annual page extent of 1200. If the journal has 10 editors and each editor accepts *just one more paper over the entire year,* that is equivalent to a whole extra issue, leading the journal to be 8% over page budget. If the editors are each handling around 50 manuscripts annually, one paper represents only a very small part of their manuscript load, and so it is not difficult for them to make such a small adjustment to the number of manuscripts they accept. If a journal is trying to increase its Impact Factor, which virtually every journal is trying to do, introducing greater stringency for acceptance in the range where there are many manuscripts of similar quality can help achieve that objective as well as ensure the page budget is maintained. Keeping editors regularly informed of page extent will move their decision making in the right direction when necessary.

5   Editors, editorial office staff and the people involved in the production process can do various things to help stretch the existing page budget. Can page-length restrictions be made more stringent? Could some of the material be moved to the supplementary material? Could display material be arranged more effectively or be sized differently to remove empty space? These measures may result in just small savings in space per article, but can add up to quite a significant saving over a year.

6   The editorial office can act to regulate copy flow by monitoring the manuscripts that are out for revision. If a journal is short of copy, authors of manuscripts that will take a relatively short time to revise can be reminded sooner than they would normally (see this chapter, page 105) and encouraged to submit their revised manuscripts as soon as possible. But if a journal has more than enough or too much copy, the reminders can be left until as late as possible. Adherence to the allowed revision time can also be stricter so that overdue revisions have to be considered as resubmissions (see this chapter, page 105). This is a subtle process, but can often be enough to regulate copy flow very effectively. Having even-sized issues is attractive to some people, but journals shouldn't be too worried about having issues of different extents, especially if a supplement is being prepared for publication later in the year or a journal is experiencing an unexpected increase in submissions. The issues at the start of the year can be smaller, with pages kept in hand for the latter part of the year. There needs to be consultation with suppliers of services (copyediting, typesetting, and printing) to ensure this arrangement is acceptable and to allow them to adjust their schedules accordingly.

It is unethical for a decision to accept for publication to be reversed unless a problem is found with a paper, for example fraud or an ethical issue. Decisions should not be reversed just because a journal has misjudged the availability of space. Cases where this has happened have been referred to organizations that deal with ethical issues in publishing. The Committee on Publication Ethics (COPE) has ruled that editors should stand by their decisions to publish unless serious problems are found, and this is incorporated into COPE's code of conduct for editors.[2] COPE also advises that a new editor should not change a previous editor's decision to publish a paper when they take over a journal unless a serious problem is found.

Special considerations in decision making: dual-use research and the possible misuse of information

Editors should base their decisions on whether or not to publish research articles in their journals solely on the quality of the work and its suitability for their journals and its readers (see Golden Rule 8). In some scientific disciplines, however, new discoveries can be used for bad as well as for good (so-called 'dual-use' research), the prime example being the creation and use of the atomic bomb. The global threat of bioterrorism at the start of the 21st century has raised awareness that new information could be misused to create weapons of mass destruction (either physical or biological), with potentially devastating consequences for the world and all its life forms, including humans. Microbiology-related work is one of the areas particularly relevant to this, and the successful chemical synthesis of polio virus cDNA in 2002 made the synthetic creation of infectious agents a reality.[3] We are now in the era of 'synthetic biology'. Various other areas of research also give rise to concern: for example, how to make vaccines ineffective, how to increase the virulence of pathogens or make non-pathogens virulent, and how to increase the transmissibility of pathogens or increase their host range.

Amidst growing public concern, the US National Academy of Sciences (NAS) and the Center for Strategic and International Studies (CSIS) co-sponsored a public meeting early in 2003 'to bring together scientists and policy-makers to discuss whether current publication policies and practices in the life sciences could lead to the inadvertent disclosure of "sensitive" information to those who might misuse it'.[4] The day after that meeting, a group of journal editors, scientist-authors, government officials and various others met to discuss the issues and to explore possible approaches. There was general agreement that certain information might present sufficient potential risk of misuse by terrorists for publication to be inadvisable. It was also agreed that it is not really possible to formulate definitions of what should fall into that category, but that public discussion might lead to the development of a set of publication policies for journals in the life sciences. An editorial following on from the discussions at the meeting was published in *Proceedings of the National Academy of Sciences of the USA*[5] (and also appeared in *Nature* and *Science*). The four statements that arose from the discussions of that day can be briefly summarized:

1   The integrity of the scientific process must be protected by publishing manuscripts of high quality and in sufficient detail to allow reproducibility and so independent verification.

2   Research in the fields that are open to potential abuse by bioterrorists is critical in the quest to combat terrorism. Safety and security issues in papers submitted for publication must be identified and be dealt with responsibly and effectively.

3   Scientists and journals need to consider the appropriate level and design of processes for the effective review of papers that raise security issues. Some journals already have such procedures and could be used as models.

4   In some cases, editors may decide that the potential harm of publication outweighs the potential benefits. In such cases, papers should be modified or not published. Journals and scientific societies can play an important role in encouraging investigators to communicate results in ways that maximize benefit and minimize risk of abuse.

Since that meeting in 2003, several leading journals have introduced procedures to deal with the issue of bioterrorism, and have a list of advisors they can consult on biosecurity issues. There are concerns, though, that procedures in some journals may not be adequate, as editors are not necessarily well qualified to make such judgements. Formal and comprehensive guidelines are therefore needed. However, the general feeling is that decisions on whether results are published should be in the hands of the scientific community and not dictated by government policy, and so an ongoing discussion between journal editors and national security experts is desirable.

Beware!

Be alert to the possibility that submitted research may have potentially harmful as well as beneficial applications.

Several countries are looking into introducing guidelines or mechanisms to guard against the misuse of research and to provide a source of information for scientists and editors to help them assess the balance of risk against benefit. Most recognize the importance of developing a system that will allow basic research to move forward without hindrance while enabling potentially harmful research to be identified. For example, in the USA, the National Science Advisory Board for Biosecurity (NSABB; www.biosecurityboard.gov) was created in 2004 to advise all Federal departments and agencies conducting or supporting life sciences research that could be of dual use. One of its aims is to work with journal editors and publishing communities to ensure the development of guidelines for publication of potentially sensitive research. In the UK, the House of Commons Science and Technology Select Committee held an enquiry in 2003 into the scientific responses to terrorism and recommended following the US example and providing funds for research on how to handle potential terrorist threats.[6] The Committee also called for the introduction of an ethical code of conduct for scientists on the potential misuse of research, and urged learned societies and the research councils to take the lead in this. The Royal Society and Wellcome Trust co-sponsored a meeting at the Royal Society in 2004 on reducing the potential for the misuse of life sciences research.[7] This was attended by academics, government and industry scientists, representatives of funding agencies and learned societies, publishers, science journalists and government policy makers. It was concluded that: preventing the publication of basic research would not prevent the misuse of advances in the life sciences (and would actually hinder work with beneficial applications); self-governance by the scientific community rather than new legislation was preferred; the scientific community should take the lead in formulating codes of conduct; and education and awareness-raising training were needed for scientists at all levels.

As currently there are no formal guidelines for dealing with the publication of dual-use research, it is the responsibility of editors of journals with content that may fall into the dual-use category to keep abreast of developments and to introduce appropriate measures and procedures into their journals. The issues should be discussed amongst the editors, and journal policy formulated and made clear to the journal's community. Some editors will already have set up monitoring groups and a network of external expert advisors. Those that have not, or who are new to their positions, should make it their duty to find out about the latest developments. Checking the websites of societies and journals for whom dual-use research is a significant problem is a good start, for example that of the American Society for Microbiology (http://journals.asm.org/), where there are policy guidelines and codes of ethics that refer to biological weapons. Editors should identify manuscripts that contain potential dual-use research and their review should be especially rigorous and take into account the special circumstances. A broader spectrum of reviewers than usual needs to be used, and some reviewers should be asked for their opinion on the level of potential risk as well as for an assessment of the scientific merits of the work. It is ultimately, however, as in the case of all papers, the responsibility of the editor after appropriate consultation to make the decision on whether to publish or not, and if to publish, whether any modifications are needed to minimize the risk or dangers of inappropriate use.

References

1   Editorial. (2003). Coping with peer rejection. *Nature*, **425**, 645.
2   Committee on Publication Ethics (COPE). A code of conduct for editors of biomedical journals. www.publicationethics.org.uk/guidelines/code (accessed 3 March 2006).
3   Cello, J., Paul, A. V. and Wimmer, E. (2002). Chemical synthesis of poliovirus cDNA: generation of infectious virus in the absence of natural template. *Science*, **297**, 1016–1018.
4   Cozzarelli, N. R. (2003). *PNAS* policy on publication of sensitive material in the life sciences. *Proceedings of the National Academy of Sciences of the USA,* **100**, 1463.
5   Journal Editors and Authors Group. (2003). Uncensored exchange of scientific results. *Proceedings of the National Academy of Sciences of the USA,* **100**, 1464.
6   House of Commons. (2003). The scientific response to terrorism. Eighth report of session 2002–03. HC415. The Stationery Office, London. www.publications.parliament.uk/pa/cm200203/cmselect/cmsctech/415/41502.htm.
7   Do no harm: reducing the potential for the misuse of life science research. Report of a Royal Society–Wellcome Trust meeting held at the Royal Society, 7 October 2004. RS PolicyDoc:29/04. www.royalsoc.ac.uk/displaypagedoc.asp?id=10360.

6 Moving to online submission and review

The online submission and review of manuscripts became feasible only in the middle of the last decade of the 20th century. By then, the Internet and email had entered the public domain, and rapid electronic communication and the ability to access and transfer large files electronically became a reality. Since the late 1990s, various web-based online submission systems have appeared on the market and significant numbers of journals, especially in science and particularly in the biomedical sciences, have moved away from their old paper-based systems to these web-based ones. Many of those journals that have not yet made the move are considering it. Any that are not, need to, because as more authors and reviewers become used to submitting and working online, and experience the benefits of doing so, they will begin to expect it of journals – this is already happening. Those journals that remain paper-based may find themselves at a distinct disadvantage.

There is much anecdotal information around about the usage of online systems but very little hard evidence. In an attempt to address this, an extensive survey on online submission and peer review was carried out for the Association of Learned and Professional Society Publishers (ALPSP) in early 2005.[1] The key objectives were, firstly, to present an overview of the systems then available (10 main ones), in a way that would be of practical use to those contemplating adoption or a switch of system, and secondly, to survey editors, authors, reviewers and publishers to find out about their experiences and to establish the reactions of the end users and effects on journals. Ten thousand recent authors were selected at random from the ISI Web of Knowledge database and invited to take part in the survey to give their experiences and opinions as both authors and reviewers. The participants came from a broad range of disciplines, but biomedical subjects were the most prevalent and the whole sample was predominantly STM (scientific, technical and medical). Of the nearly 2500 journals represented by the publishers who took part, only just over a third were using online submission systems. However, the publishers expected the great majority of their journals to be using online systems at some stage in the future. Editors who were users of online systems were recruited by the publishers who had volunteered to help with the study. The survey contains much very interesting and useful information, and readers are referred to that for the complete picture[1] and to a summary article.[2] Some of the main findings that will be of relevance and helpful to those thinking of moving to online submission and review are mentioned in this chapter.

The basic peer-review process as described in Chapter 2 (page 9) is the same for online working as for paper working. The guidelines for good practice given in this book still apply; good practice is system and business-model independent. There are, however, certain considerations that are specific to online working and these have been covered in earlier chapters. This chapter is mainly concerned with the move to online working – how to set about this and how to achieve an implementation that will be successful and beneficial both to the journal and to all the parties involved. Online systems do not operate in a vacuum or function independently. Human intervention and activity are very important and they are what distinguish a successful system from a passable one.

How do you choose an online system?

To a small publisher or journal wanting to move to an online system, this may seem like a daunting task. It may even be difficult to know where to start. As always, it's helpful to talk to others who have already taken the step. Phone journal editors or journal staff, and perhaps arrange to visit their offices to see their systems in operation. All the vendors of the main online systems will be keen to demonstrate their products and discuss how their systems can be used or adapted to individual requirements. But beware, as since they will be trying to sell you their own systems, they won't be forthcoming in bringing to your attention any problem areas or deficiencies. Ask detailed and penetrating questions (see Box 6.1). If you're not happy with the responses you get, say so and pursue your points until you are, or check them out with existing users. Many of the systems now offer a similar set of features, but some do not, and may not have features that might be important to you. You need to find out what they can't do, as well as what they can do. The largest systems in 2005 each had over one million users and were dealing with around 30,000 submissions a month.[1] The vendors therefore have a great deal of experience and are familiar with the varying requirements of different customers and will be able to advise new clients on what they might need.

Some publishers and journals may be concerned that moving to an online system will be too expensive for them. However, it is a highly competitive market and potential customers should not be afraid to negotiate costs, and also to establish some safeguards on pricing, for example to cover the situation where they may experience a large increase in submissions following their move online. It may also be possible to negotiate low or reduced charges for manuscripts rejected without review. So journals and publishers should not be put off looking into the possibility of moving to an online system just because they feel it will be beyond their means.

When publishers were asked in the 2005 survey about the most important features to take into account when choosing a potential system they gave, in order of importance: overall ease of use, customizability, ability to import data from existing systems, closest match to existing editorial workflow, and having a track record of regular and frequent updates and new releases. When asked about the most important features of the vendor, these were, in order of importance (but all were

Box 6.1 Questions to ask the service providers of online submission and review systems*

- How many publishers (with examples) use the system?
- How many journals (with examples) use the system?
- How many users are there?
- How many submissions are there per month?
- What are the pricing arrangements? Is there an initial set-up fee and annual licence, or a one-off licence?
- What is the fee per manuscript? Are there any reductions for manuscripts rejected without review?
- What are the support charges?
- Which file formats are accepted?
- Is uploading of supplemental files allowed?
- What reporting is available (preset, customized, user-defined)?
- What customer service and support are available?
- Which features are available? Are the following?

  - a hosted service (i.e. by an application service provider, ASP, who provides an outsourced service which includes hosting of the application) or licensed software
  - configurable workflows (i.e. without the need for custom programming to adapt to different workflows)
  - automatic email generation with personalization option
  - version management and control
  - automated delivery of files and metadata to production systems and suppliers
  - support for invited articles
  - support for ecommerce and billing
  - reference linking
  - reference checking and reformatting to house style
  - integration with external databases
  - image quality checking on submission
  - proxy submission
  - online copyright assignment.

* Based on categories covered in the ALPSP survey of online submission and peer-review systems.[1]

considered very important by the majority): that the vendor demonstrated under-standing of the peer-review process, long-term viability of the vendor, and support.

It is important to many publishers and journals that the commercially available online systems can reasonably adapt to their optimized workflows rather than them needing to change these significantly so as to fit those developed by the vendors. The systems therefore need to be configurable and this should be checked with the vendors when investigating their particular systems. There are various possibilities, ranging from workflows that are fully configurable to those that are not at all. In between, and most common, there will be a choice of workflows that can be con-figured to various extents. However, extensive custom programming rather than permitted configuration is not a good idea as it will make it difficult for vendors to update these systems smoothly to new versions. Journals should therefore think very carefully before insisting on a large number of additional features or changes. The goal shouldn't be to attempt to find an exact match to the existing workflow or to move everything from a paper-based system across to an online one; rather it should be to optimize the workflow. Moving to online working should be an opportunity to evaluate all systems and procedures and decide which to keep, which to modify, and which can be dispensed with. Many of the larger vendors now have vast experience of many different types and sizes of journals and so know the sorts of things that work and those that don't; they can therefore advise potential customers on this. It is interesting to note that the feature that was considered most important in the vendor was an understanding of the peer-review process. The importance of this should not be underestimated – it will make the move to online working easier and more successful, and guard against a journal ending up with a system that is not right for it.

Beware!

Do not ask for extensive customized programming when adopting an online site.

How to prepare to move to online working

The secret of a successful move to online working is rigorous planning and organ-ization, making sure everyone is on board, preparing adequately and not under-estimating how long it will all take. The ALPSP survey found that the average time from deciding to move online to going live was nearly 16 months: 8.4 months for planning, specification and system selection, and 7.3 months for implementation. Considerable forward planning is therefore needed. Once you've investigated the market, sought advice, interrogated various vendors, viewed their demo systems, decided on the right system for you and the contractual arrangements have been agreed, you need to work out how that online system is best going to work for you and how you are going to implement it. This should be done in close discussion with the vendor. Ideally, one person should act as the co-ordinator and contact person with the vendor. Larger publishers or journals may be able to assign a full-time individual to project manage and troubleshoot, especially if a number of journals are being moved over at the same time. Whatever the organization of this, it is vital that all parties are involved from the start, kept regularly updated, and brought into the decision making when appropriate. Everyone needs to be on board and working together to achieve a successful move. What steps need to be taken next?

Beware!

Don't underestimate how long it will take to move to online working.

Evaluation of current workflow and responsibilities

Put down on paper your current workflow, i.e. the path a manuscript takes from submission until final decision, and add the people involved at each stage, noting who does what and who makes the decisions – indicate every point in the process that a decision of any sort needs to be made. Take a close look at the final scheme and analyze each step, asking whether it is necessary, and if it is, whether it is the best way to deal with things and whether the appropriate people are involved. You may be using a system that originated years ago with a distant and long-gone editor and which, although it has evolved, hasn't really changed in basic structure and may have redundant or overcomplicated steps in it.

The number of layers of editorial responsibility will probably reflect the size and complexity of your journal. For very small journals there may be only one level, with the editor-in-chief receiving manuscripts, contacting reviewers, making decisions and communicating these to the authors. For larger journals, different editors and editorial office staff will be involved. Staff will probably receive manuscripts and not pass them on to editors until they have been checked for completeness and quality and all deficiencies and problems have been dealt with (see Chapter 3, page 32). The manuscripts may then go through a single main editor or directly to a group of subject or specialist editors (see Chapter 3, page 35). These editors may then pass manuscripts on to another group. A general key point to remember is not to have an overly complex workflow or to assume you need to have all the functionality on offer with online systems. An overcomplicated system may end up just confusing your various user groups, perhaps even to the extent that they refuse to work online.

Beware!

An overcomplicated online system may confuse or alienate the users.

It is critical to work out who is going to have decision-making powers on the site and who will be contacting authors and reviewers. In theory, editors can deal with both groups directly and can also add information to the database and alter existing data. Some journals may therefore feel this is an ideal opportunity to move some work from the administrative personnel to the editors. However, although this may seem superficially attractive, various factors need to be considered to avoid ending up with negative consequences. For example, if you have a hundred 'editors' on your journal and they are all given this opportunity, how are you going to guard against degeneration of the database, overloading of some reviewers, and inconsistency in decision making? There is a lot to be said for leaving database changes, such as adding or deleting users, to the editorial office staff. This will ensure this is always done systematically and accurately. Similarly, decision making is better left to relatively few individuals (see Chapter 5, page 87). Some, indeed perhaps many, editors will also not want to take on the role of direct communication with reviewers and authors and the increased administrative load this will bring. They may soon become overloaded and begin to question whether they want to carry on, as the job is not what they signed up for, and they are no longer able to concentrate on the scholarly aspects of the work as much as they need and want to. Direct communication with authors and reviewers means these people will also be able very easily to reply directly to the editor, perhaps asking questions that are normally dealt with by editorial staff, and this again will increase their workload. It will also risk queries being forgotten and so left unanswered, especially if the editorial staff aren't aware of them.

Beware!

Take care not to overburden editors with administrative tasks just because the functionality is there.

It is very important to consider all the editorial staff members and how they may feel once the decision to move to online working has been made. They will naturally be concerned, as major change is always difficult. Their roles will undoubtedly alter. They will have concerns, perhaps never voiced, not only about this, but also as to whether their jobs will be downsized or even disappear. For some organizations, some jobs may disappear and this needs to be discussed with the staff. Journals should, however, bear in mind that there have to be enough staff to cope with the increased submissions that many journals experience after moving to online submission (see this chapter, page 132). In other organizations, the opportunity is there for staff to be moved to different roles or to be given different responsibilities. In these cases, staff need to be reassured and shown that, although their jobs will change, it will be for the better. They will become much more 'customer focused' and interesting, as some time previously taken up with sheer clerical work and administrative support will no longer need to be devoted to that and more time can be spent on value-added activities. It is an opportunity for personal, as well as journal, growth and development. Some staff may be concerned that their computer skills will not be sufficient. Again, these concerns might not be voiced because the individuals may feel they are putting themselves in a vulnerable position by admitting to them. Any skills deficiencies need therefore to be diplomatically identified and appropriate training arranged well in advance. It is very important that all the staff feel comfortable and confident enough to take on the new technology.

Specification design for the journal

Once you have decided on the best workflow from the journal's point of view, enter into discussion with the vendor to see how well it fits with any of their standard workflows. It may already exist, or can be achieved with just a small amount of configuration. The vendor's input may cause you to rethink some of your workflow. Together, you will be able to devise the best option for your specific journal and community. The vendor will then provide you with a specification form for you to fill in about other requirements and preferences. A few years ago, these forms were quite basic and required a lot of thought on the part of the customer. As vendors have dealt with more and varied customers, the forms have become far more extensive and cover many things that perhaps some journals won't even think of. But don't take up all the options just because they're there. Ask yourself if you need them and whether they'll add value to your online site. Some may well be inappropriate. Critical analysis is needed here. Your site will then enter development by the vendor to get it ready for you to test. During that time you need to be getting other things done and people prepared.

Questions and text options on the site

Some of the wording on your online site is likely to be standard across all journals using the same online system, but with many systems you will, to varying degrees, be able to customize some of the wording and add a number of journal-specific text additions and questions. Opportunities for the latter may be limited, so use the options wisely. For example, many online systems can be set up so that authors have to tick various boxes before the system will allow them to proceed with a submission. This is a useful place to put statements related to authorship and journal policy, so that it is clear to authors early on what these requirements are and that they need to agree to abide by them for the journal to consider their paper. No author can then complain at a future date that they were not aware of journal policy or requirements. Some examples of useful statements that authors must confirm before they can submit a manuscript are given in Box 6.2. The first statement is an all-encompassing one, assuming a journal has made clear its policy requirements in its Instructions for Authors. If it hasn't, it should (see Chapter 3, page 26).

> ## Box 6.2 Examples of statements for authors to confirm before they can submit a manuscript
>
> ☐ This manuscript has been prepared in accordance with the journal's Instructions for Authors and follows all the journal's policy requirements.
>
> ☐ All authors are listed, each author has participated sufficiently in the work to take public responsibility for the content or part of it, and each author has approved the final version of the manuscript.
>
> ☐ No part of the manuscript has been, or will be, published elsewhere nor is under consideration for publication elsewhere.
>
> ☐ All potential competing interests, whether financial, commercial, scholarly or personal, have been disclosed.

Journal-specific text can also be added on some systems, and can be highlighted in a different colour and/or presented in bold font or capitals. This can be very useful, especially if it is clear that authors are having problems at certain places in the submission process or omitting to fill in certain important boxes. Adding simple instructions at such places can make everyone's lives much easier and avoid considerable frustration for authors. For example, on some systems the final 'submit' button may be at the bottom of the immediately visible final page screen and authors need to scroll down to see it. If they don't press this button, the manuscript won't be submitted even though an author may think it has been. Lack of an acknowledgement or reference number isn't enough in some cases to alert authors that they haven't completed the submission process, and it may only be a month or two later, when they contact the journal to find out about the status of their manuscript, that it becomes apparent what has happened. Submission can then be completed, but valuable time will have been lost for the author. Adding a line of text to tell authors they need to scroll down and press the 'submit' button helps avoid these incomplete submissions.

> ## Beware!
>
> Authors may fail to press the 'submit' button when attempting to submit their manuscripts online.

Preparation of editorial correspondence for the online site

Editorial letters need to be composed for every stage of the submission, review and decision process. The skeleton text of these needs to be added to your online site and the manuscript details will, in many systems, be added automatically when the correspondence is generated. Do not underestimate how long it will take not only to prepare the correspondence, but also to input it and include the various parameters, such as which standard details are to be included in each letter and who is to be included in the mailing. It won't in most cases be appropriate to use exactly the same correspondence that is being used in the paper-based system; it will need to be adapted. The letters need to include information about online-specific matters, such as how to submit a revised manuscript and where to put responses to reviewers' comments, as well as details of the decision and editorial process. Appendix II gives some examples of standard editorial letters.

It can be very helpful and effective to include brief instructions at the bottom of some correspondence to jog users' memories as to what to do next. Editors may forget exactly what to do and where or how to submit their communications and decisions, especially if there are times when they are not very active. They don't really want to have to keep referring to long pages of instructions they may have been sent some time ago (assuming they can still find these and they are up to date) and will almost certainly appreciate brief reminders. They can always ignore the instructions if they don't need them. Appendix II (page 261) gives an example of the sort of message that can be sent to editors when new manuscripts are assigned to them. The instructions included are concise but comprehensive. Editorial staff should be sure to check all the master correspondence regularly and update it as necessary.

It will take time to decide and input who is to be copied in on each piece of correspondence generated by the system. After the initial excitement of going live, new users will probably find that they have copied in too many people and nearly everyone in the office is being copied in on everything. The novelty will soon wear off, as, in some cases, hundreds of emails start arriving in people's in-boxes. There will be information overload, and the risk that important messages will be lost in the flood of standard, for-your-information type of messages. The copy facility is an invaluable one, however, and should be used to alert appropriate people that an action has occurred or that a review has been received – so that, for example, an editor monitoring other editors' decision making knows when their decisions have been submitted and are ready to be dealt with and communicated to the authors.

It is also good practice for all users to blind-copy themselves in on the messages they send via online systems. This ensures the correspondence has been sent and received – it can be immediately deleted, or moved to another electronic folder for reference if this is how an individual likes to work and keep track of their own areas of activity. On those occasions when correspondence is not received (which is relatively rare nowadays with modern systems), the service provider needs to be contacted and the problem reported and sorted out.

Health and safety issues

It is inevitable that moving to an online system means that users will spend most of their editorial time working at their computers on a screen. For the editors, authors and reviewers it will be an intermittent activity. For the editorial office staff, however, it may mean that they will have to spend their whole working day at their terminals. Normal health and safety guidelines should be observed for break times and staff should be encouraged to follow them. Many editorial staff are very dedicated individuals and may tend to skip these – their managers should be alert to this and periodically remind them, and also encourage staff to plan their work days so as to fit in some non-screen activities.

> **Beware!**
>
> Current physical working arrangements and infrastructure may be inadequate and unhealthy for online working.

The desks and chairs used by staff need to be looked at and if they do not offer adequate support for online working they need to be changed. Fully adjustable chairs with proper lumbar support and curved desks that offer elbow support are advisable. Working on online systems is not like doing regular keyboard work; there is much greater use of the mouse, with both left and right clicks, and a considerable amount of scrolling. A good mouse mat with a gel-filled wrist support is invaluable. Be very alert to any physical problems that may arise and attend to them promptly.

Training and support

The importance of appropriate and thorough training before going live online should be fully recognized. Fully trained users who know what they're doing, who are familiar with all aspects of the system and know how to assist other users will help make your move to online working a success. Inadequately prepared staff may find it very hard to adapt to the new method of working and may very soon feel overwhelmed and lost. Some may leave. Morale and confidence will plummet. When publishers were asked in the ALPSP survey what, with hindsight, they would have done differently in implementation, many of the comments were about making sure everyone (staff and editors) had sufficient training to enable them to become familiar with the system to get a better understanding of the whole process, and making sure that enough time was allowed for this.[1] So start the training phase early – do not leave it until just before going live on the system. Once the vendor has provided you with your test site, give everyone in the editorial office the opportunity to play around on it, to submit dummy manuscripts and take them through the whole editorial process. Encourage staff to take on proxy roles (i.e. acting as authors, reviewers and editors) so they become familiar with the whole system and know what can and can't be done, and also so they can experience the potential problems various users may encounter and can work out how to resolve them. Once your online site is live, it will no longer be possible to play around on it; all actions will have consequences and remain on the site as a permanent record and will be included in the generation of statistics and in reporting.

You may find that different members of staff have very different aptitudes for online working. Some may find it difficult to adapt and may even feel threatened if more junior members seem to be coping very easily and have no problems navigating and working on the site. Arrange individual tuition between members, or let certain people be group leaders in regular training sessions devoted to specific areas or problems. The service provider will also arrange training and support as part of the package, the extent of which will vary. Use all you are entitled to. If you feel you need extra to that included, make arrangements for this. It will be worth the investment.

Draw up detailed step-by-step instructions for everything on the site – for all stages of a manuscript's flow through the system and for all users. Make these comprehensive but concise. Incorporate screen shots wherever these help illustrate a point. Draw up special sets of instructions for the editors. Don't overwhelm them with information and don't include information on things or steps they will never need to use. Organize the instructions into well-labelled sections so the editors can easily and quickly find what they need to when they're dealing with different manuscript stages. Editorial office staff may by this time be very familiar with the functionality and features of the system and may forget that not everything is immediately obvious to a new user. Don't assume knowledge of even the simplest features, especially as members of most editorial teams vary significantly in their aptitude for computer-related matters. The importance of training is reinforced by the ALPSP survey: nearly half of the editors did not feel that online systems were simple to use without training, but three-quarters felt that the systems were simple to use after appropriate training.[1]

An important finding of the ALPSP survey was that editors who did not receive good training and support during the transition to an online system ('unsupported editors') were much more likely to hold negative views about the systems and their ease of use than editors who did receive good support. Publishers also reported that getting buy-in from editors was important in successful implementation. The areas that gave most problems in implementation were editors who were reluctant, conservative or computer illiterate, and just managing the change and the training requirements. A significant proportion of the publishers (over a quarter) reported that some of their editors and reviewers were so unhappy about the changes (or were unable to adapt) that they left the board or declined to review further. The numbers reported were small (and referred mostly to reviewers), but journals should work hard to try to minimize them as these individuals may leave with a very negative, and perhaps unjustified, opinion of the journal that they may communicate to others. In some cases, however, the reluctance to accept change, despite good training and support, may be so great that there is no choice but for journals and editors or reviewers to part company. But the large majority of editors in the ALPSP survey were very happy with online systems. Most agreed that it made it easier for them to do their jobs, and that online systems allowed them to offer an improved service to their authors. Just over half felt that they were better able to support their reviewers. A large majority (over three-quarters) felt that lack of an online system would be a negative factor in considering offers of other editorial positions. Again, this is an important statistic to be taken on board by journals hesitating about moving to online submission.

Data transfer

If you are moving from a totally paper-based system, you will not of course be able to transfer any electronic data to the online system. All you will probably have time to enter will be details of the editorial office staff and other regular users, such as editors, and perhaps those of your most active reviewers. People who are no longer active as reviewers, because they have requested not to be, because the journal no longer wishes to use them for any reason, or because they are retired or deceased, should also be added so that they are not contacted in error (see Chapter 4, page 48). The rest can be added as they need to be once your online site is live.

If you are already on an electronic system of some kind, or have a reviewer database, you will in most cases be able to transfer that information over to your online system. Seek advice early on from your service provider and prepare the data files as they recommend. Pass them over to them in good time so that the data can be transferred successfully to the online system before launch. Don't leave it so late that there will be no time to sort out problems, but equally don't do it so early that the database will end up incomplete because the most recent information will be missing. Agree on a realistic timing schedule with your service provider, who should be able to advise you on all aspects. You won't in most cases be able to transfer data on manuscripts submitted and dealt with on your old system, so make arrangements to maintain access to records by storing them on a disk drive or elsewhere electronically. Careful thought should be put into how much of your existing electronic database you want to move over. There may be people on it you haven't used for years. It's an opportunity to cull the database and take over only those people who have been active, say, in the past 3 to 5 years. Again, it's a good idea to move over details of the individuals you don't want to contact for reviewing in the future, those who have requested not to review, and those who have become inactive due to retirement or death so that you don't accidentally try to use them as reviewers. The remaining names can be stored electronically locally for possible future reference. Once live, the online database will build very rapidly, as everyone who submits, edits or reviews for the journal will be added as a matter of course.

The launch and transition period

You will need to decide how to go about announcing the move to online submission and review, and how and when to launch it. You need to be totally prepared but, as in most things in life, it is unrealistic to wait for the point when everything is 100% ready. Some staff may still feel a bit insecure and worry about the minutiae.

Someone needs to take control and make the decision on when the launch date should be. At that time the online system should be working well, all the correspondence letters should be mounted, everyone should know what they are doing, and support measures should be in place. An announcement about the online system needs to be made. The journal's website is by far the most important place for authors to find out about the move to online submission. Details should also be added to the paper copy of the journal and, if the journal is a society-affiliated one, to the society's website and newsletter. Many publishers will be willing to send out email alerts to potential users for their journals; in this case, make sure the message contains a direct link to the online site so that potential users can click through without any effort. The journal community also needs to be advised whether all submissions are to be online from launch date or if both online and paper submissions will be considered. Some service providers are now recommending that journals ask for all submissions to be done online from the date of introduction. The decision will depend to some extent on the community served by the journal, as different communities will vary in their expectations and response. Editors and editorial staff are therefore the best placed to decide on this. They should also seek advice and guidance from both their publisher, if they have one, and their service provider, as they will be able to give them information on similar journals and on likely outcomes.

If paper submissions are to be allowed, you need to decide how to deal with these. You can continue to run electronic and paper submissions side-by-side, but this may become very labour intensive and may slow advancement of the online system. Confusion can also arise if a new workflow is being introduced with the online system but the old one is being maintained for paper submissions. The online database will also not be complete for that year, nor will the statistics and reporting, and dual systems will need to be run and the results amalgamated to produce the annual reports. One option is for editorial staff to scan all paper submissions and then submit them onto the online system by proxy for the author. After this initial extra effort, the manuscripts can be handled in exactly the same way as online submissions. Alternatively, authors can be asked to provide electronic files and these can be submitted by the editorial staff.

Uptake of the online option if there is a choice will vary from community to community, but as so many authors nowadays are used to submitting online (in science in particular, and indeed in some disciplines they expect to be able to do this), uptake may be very rapid and so it will be a steep learning curve for the editorial office staff. A time will come when journals who have continued to allow paper submissions will decide to make online submission mandatory except in exceptional circumstances, and with agreement from the editorial office, when online submission is not possible for one reason or another.

When the day of launch arrives it's an important occasion. It will have been a long, and perhaps difficult, journey to reach this point. It will undoubtedly have involved a lot of extra time and effort by many people. There may have been low points, and even tears, along the way. Have a bottle of champagne ready and celebrate when the first online manuscript comes in – it's an exciting moment!

What to expect after going live online

Journals that moved to online submission in the early days of online system availability didn't know what to expect. Many experienced increases in manuscript submission. As these were unexpected, they couldn't be planned for. Data are now, however, available on the impact of online systems on submission numbers, the quality of submissions, and the geographical origin of submissions.[1]

Impact on submission numbers

Most journals can expect a reasonable increase in number of submissions; most commonly around 25%. This needs to be taken into account in planning. More editors may need to be recruited to cope with the extra manuscripts to avoid handling delays and overloading the existing editors. The page budget may need to be increased and/or the threshold for acceptance changed to avoid building up a backlog of papers awaiting publication and therefore publication delays, which has been a problem for some journals. Editorial office staff need to be alerted to the likely increase in submissions. The reduced clerical and administrative time associated with moving to online working frees up time to deal with the extra submissions.

Impact on the quality of submissions

Increased numbers of submissions are not necessarily a healthy consequence, especially if the quality of the extra submissions is low or the manuscripts fall outside the scope of the journal. This will just create extra work and eat into resources without bringing any benefits. Journals should therefore monitor closely the effects on submission quality and scope as well as just number, as the policy changes needed to deal with different types of increases will be different. If increased numbers of out-of-scope manuscripts are being submitted, the policy guidelines for authors may need to be clarified. If increased numbers of lower-quality manuscripts are a problem, the editors should be alerted to this and editorial policy discussed to ensure consistency in acceptance, both for external review and for publication after review.

Impact on geographical origin of submissions

The effect of online submission on the geographical origin of manuscripts is difficult to predict, and will vary from journal to journal. It is likely, however, that submissions will become more international.[1] This is partly due to the greater ease and speed of submission and partly due to the lower costs (photocopying, figure preparation, postage) for authors. For journals trying to extend their reach, this will be a welcome consequence.

Impact on reviewing and administration times

Journals that move to online submission and review can expect to see reduced reviewing times, most commonly around 25%; administration times will also be reduced, possibly by around a third.[1] Both of these are very positive features.

Problems that may be encountered and how to deal with them

Online systems are never closed

Online systems are never turned off – they are like a running tap. Staff who are used to receiving paper manuscripts at regular post times and clearing their work before leaving each day (and don't feel comfortable unless they can achieve this) will need to adapt psychologically to this. Emails and submissions may sometimes seem to pour in at the end of the working day as editors, authors and reviewers in different time zones start working. They will need to be ignored and left until the next day. Similarly, the number of emails and new submissions waiting on Monday mornings may be daunting – many authors will have used their 'spare' time at weekends to devote to manuscript preparation and submission. Staff need to be helped to adapt and not feel overwhelmed. They need to learn to order their work systematically, making sure the most important things are dealt with in a timely fashion and nothing is overlooked, but that work does have to left until another day.

Users will grumble

Various users will complain and grumble about the system. Some of these grumbles may be genuine and will need sorting out. If a number of users are experiencing similar problems with some aspects of the system, notify the service provider. They may be able to sort it out immediately, or they may, if enough users are affected, add it to their development list. Be aware that some users will grumble away regularly at various things, or even at a great many things, but the problem may lie more with them than with the system itself. They may also not be able to appreciate that it is not possible to change some things at will on commercial systems. Some complaints may arise because users aren't fully prepared or supported, not because of any problem with or malfunction of the system. This needs to be addressed and appropriate training provided to help increase familiarity with the system, and clear instructions sent that can act as reminders of the various processes and steps when needed.

Users needing extra support

All journals will have a number of users who will need more support than most. They may also very likely need support every time they act in a different role. Be prepared to be patient and help them to gain confidence. Effort put into the early stages of usage will pay off in the long run. Don't ever make users who are very willing but clearly not the most computer literate of people feel stupid or that they are the only ones who have problems. You don't want to risk either that they will hide their inadequacies and not use the site to its full capacity, or that they will stop working for the journal and a valuable member of the community will therefore be lost to it.

Editors not using the online system properly

Editorial staff cannot see what editors are seeing and doing in detail. Some may not be using the online system properly and making things much more difficult for themselves than necessary. It's a good idea to run a 'master-class' session at an editorial meeting, where you can access your online site and show editors how the screen should look optimally and how to navigate the site and use the system's various features. Some problems may be due to something as simple as an editor using an inadequate Internet browser or too large a font size and so having constantly to scroll up and down and side to side to view all the information on the screen. They may also not know how best to print out reviews and manuscripts, and hints such as 'print a few pages at a time' for the latter can help them enormously.

The need to work 'offline' occasionally

All manuscript work should ideally be done via the online system so that the information builds up and is available to everyone associated with that manuscript editorially and so that it can be referred to in the future. Occasionally, however, it will be more appropriate to communicate off the system – for example, if sensitive issues regarding the choice of potential editor are involved, or if a manuscript and any of the individuals associated with it are involved in investigations of alleged or suspected misconduct (see Chapter 9, page 185). If it is felt after an offline session that information is appropriate for the online system and viewing by all who will have access to that manuscript, it should be mounted with an appropriate summary, heading and date(s).

Users not using the online system to communicate or not mounting all relevant information

Editorial office staff will very soon get into the good habit of mounting all relevant information about manuscripts on the online system and communicating via it so that the information registers and becomes part of the permanent record. They will also remember to add concise and informative titles to their communications so that the correspondence record is self-explanatory and it is very easy for anyone to locate specific information or correspondence. Editors will sometimes forget to do the former, and will certainly forget to do the latter much of the time! Show new editors how to communicate via the online system, i.e. by using the 'hot link' email facility, for those systems that support this (i.e. an email box opens up for a person by clicking on their name and the message when sent registers for that manuscript on the online system), and remember to tell them that they can mount any information on the system for any manuscript by emailing themselves. Details of phone calls, conversations, and so on, can be mounted in this way. Some of the information may be more appropriate for the notes screen, so again, advise editors on how to use this facility. A number may be nervous about using it.

Responses from reviewers

One of the main grumbles of reviewers is having to log onto online systems just to respond – either positively or negatively – to a request to review. Many would much rather do this by email, and many will actually do this whatever the journal instructs them to do. If journals don't allow their reviewers to accept an invitation by email, they may be alienating a number and losing out on getting a response because reviewers don't feel they should be required to expend a lot of effort just to say 'yes' or 'no'. Email replies are just as easy to deal with, and important information can be added that it might not be possible to include in a response given via the online system. Responses can easily be added to the appropriate manuscript screen by including them in a special dedicated reviewer section of the notes screen (see Chapter 4, page 44, and Box 4.1).

Reviews submitted for the wrong manuscript or comments in the wrong place, and editors submitting the wrong documents or working on the wrong manuscript screen

Problems with reviewers submitting reviews for the wrong manuscript or comments in the wrong place have already been mentioned (see Chapter 4, page 75), as have problems with editors submitting the wrong documents, working on the wrong manuscript screen, or possibly emailing the wrong person (see Chapter 5, page 100). Reinforce the '3N' Rule (see Chapter 2, page 22), so that all editors and staff get into the habit of checking manuscript Number, Notes screen, and Name every time they do any work on a manuscript or communicate with anyone via the online system.

Other editors or people not associated with a manuscript needing to see material

Occasionally, information about a manuscript and possibly the manuscript itself may need to be shared with other editors or individuals not associated with the manuscript. In most online systems they will not have access to the manuscript and its details as that is usually restricted to people directly involved with the manuscript. One solution is to email them a message and any information that can be copied and pasted via the online system and then send a portable document format (PDF) file of the manuscript (and of any other material that is better sent as an attachment) by regular email if that can't be done via the online system. Their involvement is therefore registered on the online system but no complicated, and perhaps misleading, arrangements (such as making them temporarily the 'editor' or 'reviewer' for that manuscript) need to be put in place to get them access to the manuscript and relevant details. It may also not be appropriate for them to view all the information and correspondence about the manuscript that would be available to anyone in the 'editor' role, so this safeguards against that.

Submissions from the editor-in-chief

For journals with more than one editor, submissions from the general editors are fairly straightforward to deal with on online systems, as these editors will usually only be able to access those manuscripts for which they are responsible. However, an editor-in-chief will generally be able to access all manuscripts, and so if they themselves submit a manuscript they can in theory access all the information on it, including the names of the reviewers and any confidential comments they may have made to the handling editor and the correspondence leading to the editorial decision. Measures need to be put in place to prevent this. The manuscript can be submitted as usual, but for many systems it will be necessary to work offline to some degree and to ask for all confidential comments to be sent that way. For journals that do not have open review (see Chapter 4, page 42), the reviewers can be allocated general identifiers, for example be called Reviewer A, Reviewer B, and so on, and dummy accounts set up for them for that manuscript. Alternatively, the reviewers can be asked to submit their reviews by email rather than via the online system so that their anonymity is not compromised. Some reviewers may need reassurance about protection of anonymity once they realize they will be reviewing the chief editor's own manuscript. The editorial staff, or editor if it is editors who deal with reviews, can then submit the reviews on their behalf so that they will be available on the system as a record. Care should also be taken about what is put into the notes screen so that reviewer anonymity is not compromised and confidential information is not disclosed. If a journal has only a single editor of any sort, then that editor should not be submitting manuscripts to the journal, as there is no one with decision-making responsibility to handle it.

Reviewers requesting hard copy and/or refusing to work online

A small number of reviewers may absolutely refuse to deal with a submission online and request that a hard copy be sent to them because they are not prepared to pay the costs of printing out a copy of the manuscript. It is up to each journal to work out its policy on this, and whether it is prepared to send out hard copies, and if so, to whom. It may have very valued reviewers who have good reasons for requesting a hard copy and so be more willing to accommodate them. Many journals will find that it is only a very small proportion of reviewers that do this; in the ALPSP survey of online systems, only 6% of reviewers said they preferred hard copy by mail.[1]

Users adding other people's email addresses to their own accounts

The users of most online systems can go into the systems and set up their own accounts, including adding their email addresses. Occasionally a user, particularly when submitting a manuscript, may add another person's email address to their own in their own personal account. This is probably done so that that person will also receive all communications connected with the submission of that manuscript; it may, for example, be one of the co-authors or the person's secretary. If this isn't spotted, some very serious breaches of confidentiality can occur in ignorance. If person A has added person B to their account, then when person A is invited to review or sent a manuscript the communications will in some online systems also go to person B. So person B will be receiving things they should not be unless steps are taken to remove their email address from person A's account. It is good practice always to check the recipient listed at the top of an email before pressing the 'send' button. The second person's email address can then be removed at this stage. Journals may decide not to allow authors to add any other names to their user accounts and advise them of this on submission of a manuscript. If left for the duration of the review period, the second name should be removed when the review process has been completed. In the meantime, an alert notice should be added to the notes screen for the manuscript and to the notes on the individual's record so that anyone contacting that person is aware they need to delete one of the email addresses in that account.

Beware!

Some users may add other people's email addresses to their own user accounts on online systems.

A final note

Many editors and journal editorial staff may be anxious about taking the step to move to online submission and review. However, the great majority will find that this is a very positive move and can bring great benefits, as outlined in this chapter, to both the journal and to all its users. Many, indeed, will wonder how on earth they ever managed with a paper-based system! So don't be afraid to take the first step.

References

1   Mark Ware Consulting Ltd. (2005). Online Submission and Peer Review Systems: A review of currently available systems and the experiences of authors, referees, editors and publishers. ALPSP. www.alpsp.org/publications/pub13.htm.
2   Ware, M. (2005). Online submission and peer-review systems. *Learned Publishing*, **18**, 245–250.

7 Reviewers – a precious resource

Reviewers are crucial to the success of any journal. They are a precious resource and should always be treated well, and with courtesy and respect. A journal that treats its reviewers well will be rewarded with loyal reviewers who will review for the journal most times when asked, and will carry out thorough and timely reviews. These reviewers will also with time build up a very good understanding of the journal's scope and of the types and quality of papers it is trying to attract and publish. If new reviewers like how a journal operates and how they've been treated, they may be encouraged to submit their own papers to it, so the journal will also gain there. As the reviewer database is one of the places journals will look when appointing new editors – this can be very successful – it's important to build up a list of good reviewers, ranging from established members of the community to the new rising stars who may be future potential editors.[1]

Thanks and feedback to reviewers

Reviewers should be thanked for their reviews and given feedback on the outcome of the review process. Ideally, they should also be provided with all the reviewers' comments for authors (anonymous unless open peer review is used) and any other relevant material (for example if there has been an appeal; see Chapter 5, page 103). Reviewers greatly appreciate getting this feedback, but perhaps the practice isn't as widespread as it should be, as many reviewers still complain that they have no idea what has happened to manuscripts they've reviewed unless they happen to see them published in those journals and so know they were accepted, or receive them to review by, or see them published in, other journals, in which case they must have been rejected (or withdrawn).

Not only is it courteous to let reviewers know the outcome of the review process for a manuscript with which they've been involved – they have, after all, given up their time to assess it and may have invested many hours in this – it also has beneficial effects for a journal: it acts as an educative tool. Firstly, it informs reviewers of the scope and standard the journal requires for work to be published in it and so they become more familiar with those, and how they might be evolving over time; secondly, reviewers will be able to see things they may have missed in their reviews but which others have picked up. There may be points that cause them to modify their own opinions or ones with which they disagree. They may even contact the journal about this, and perhaps provide important additional information.

Letting reviewers have the outcomes and other reviewers' comments is very straightforward for journals with online systems, as these can often be set up to have this functionality. For example, the reviewers can automatically receive the outcome and reports for authors at the same time as the decision goes to the author. The process is error free, as the online system picks up the correct reports and sends them to the correct reviewers. It also does not involve any work or time on the part of the editorial staff. Certain systems allow reviewers to check online the outcomes and reviewers' reports for manuscripts they have reviewed, thus enabling them to compare their own recommendations and reviews against the final outcome and the other reviewers' comments. Some journals mount their reviewer reports on a website for access not only by reviewers, but also by others, and with the reviewers' names and the authors' response (for example, *Biology Direct*; www.biology-direct.com). The BioMed Central medical journals post the pre-publication history of each article with it, including the various versions submitted, the named reviewers' reports and the authors' responses (www.biomedcentral.com/info/about/peerreview).

The process of letting reviewers have feedback is more time-consuming for paper-based peer review because the reports need to be copied and sent by regular mail with a cover letter to each reviewer. It is also open to error, as staff will frequently be dealing with a number of manuscripts and need to make sure the right reports go to the right reviewers and that no parts of the confidential reports are included. They also have to be careful that all information that might identify the reviewers to one another, for example fax headers, has been removed. Those journals with their own stand-alone electronic systems can send outcomes and reports by email, but again, care must be taken to send the right reports to the right reviewers. For any journal without an automatic reviewer feedback system, this task will be time-consuming and so, even though it is a very valuable one, it may unfortunately be one of those jobs that falls by the wayside because of insufficient resources. All efforts should be made to try to avoid this happening.

Reviewer training

Feedback on reviewing is educative for reviewers and does constitute a sort of 'training' for them. It is especially useful to young or new reviewers who may not be very familiar with reviewing in general or with that specific journal, respectively. But how are reviewers trained to review? The quick answer is that they are not. For many scientists it will have been a case of picking things up along the way. Some may have had good mentors in their scientific careers – those who have taken the time to instruct them in the art of reviewing, and perhaps shared reviewing tasks with them in a positive and constructive way. Some will not have been so fortunate, and although they may have been involved in reviewing, it will have been more as a source of free labour and they may have been left to it without any guidance or constructive criticism, and probably without their help being acknowledged in any way. There is a growing feeling that as peer review is such a critical component of scientific communication, the responsibility for training should be taken by the community at large and should actually be a part of people's general education, to create an awareness of what it is and to help the public evaluate conflicting research claims and know what to believe in scientific reporting and what to view with scepticism.[2] Training on peer review should also form part of all courses in science, leading to more specific training for those who go on to study it at postgraduate level.

To date there has been little work done on reviewer training, but the *BMJ* (*British Medical Journal*) reported on a study in 2004 to determine the effects of training on the quality of peer review.[3] In this study, reviewers were given training either through attendance at a workshop or through a self-taught training package. Review quality was evaluated by the validated Review Quality Instrument (RQI; see Chapter 4, page 63), by the number of deliberate major errors identified, by the time taken to review, and by the proportion recommending rejection of the manuscripts. Unfortunately, the result was essentially negative, in that the small improvements found in some measures of review quality were not of editorial significance and were not in any case sustained. Training also had no impact on the time taken to review, but was associated with an increased likelihood of rejection being recommended. The authors concluded that 'very short training has only a marginal impact. We cannot, therefore, recommend use of the intervention we studied'. Despite the negative result, the *BMJ* has decided to use the intervention to provide a service to reviewers and in the hope that the training will prove useful. It has, therefore, published the learning materials online and will be holding regular half-day workshops.[4]

Many would agree that peer review is a skill that is gained by practice and not just by acquiring theoretical knowledge about how to do it. It relies on personal attributes as well as on relevant specialist knowledge. All reviewers need to have the latter to be able to assess the work in submitted manuscripts in any depth. Personal attributes determine how thorough a reviewer is, how good they are at spotting errors or flaws and at recognizing innovative research, and how conscientious they are in returning reviews in a timely way. They also determine how prepared reviewers are to spend the time needed to produce a good and comprehensive review that will be helpful to both editor and author. Exposure to people who are good reviewers and feedback from journals can help produce good reviewers. Journals can put together information packs for new reviewers that contain more than the usual guidance documents sent to all reviewers with manuscripts: for example, information on the general philosophy of the journal and the audience it is aimed at, journal policy on what is/is not acceptable for publication, which criteria manuscripts must meet, which technological restrictions will preclude acceptance, what sort of things the editor would like reviewers to comment on or watch out for, the optimum way to present reports.

Ways to recompense reviewers

Traditionally, reviewers in science have assessed manuscripts without any form of payment or recompense. They have generally considered reviewing to be a part of their jobs and the scientific process. Young scientists will inevitably feel flattered when prestigious journals start asking them to review for them, as it is recognition that they have reached a certain level of expertise and respect, and so is a reflection of their standing in the scientific community. Reviewing is a give-and-take activity. Scientists need to have their own manuscripts reviewed and will receive valuable feedback and suggestions for improvement and future directions for their research. Sometimes a reviewer's feedback will prevent them publishing prematurely and possibly making fools of themselves in public. In return, they provide the same service for other scientists. Indeed, many feel that anyone who publishes in a journal should also review for that journal, and disapprove strongly of those scientists who regularly refuse to review manuscripts but expect their own to be reviewed, and to be reviewed quickly.

Some journals offer certain rewards as incentives to get reviewers to review and to help ensure timely return of reviews. It is generally recognized that cash payments are not the ideal solution, as in most cases scientists are predominantly time-short and the relatively small payments that could be made would not make any difference to this, or to whether people will review a manuscript. There is also the danger that a payment system would impose a severe administrative burden onto already overstretched staff, and perhaps lead to increased subscription prices. Payment may also just encourage less-qualified individuals to review solely for the cash. There have been experiments with charging authors on submission to guarantee a rapid review (for example, the *Journal of Interferon & Cytokine Research*). Care must be taken in these cases that reviewer selection isn't biased to ensure that only reviewers who will be fast are used, or that completeness and thoroughness of review are sacrificed in order to meet the short deadline. There are exceptions where payment is more the norm – for example, where a general expert such as a statistician is needed, who will be much in demand and be required on a very regular basis, and so would carry a very much heavier load than other reviewers. Journals may decide that to get the quality and timeliness of review that they require, they will pay reasonable fees for such specific expert input.

Various suggestions have also been made, and some tried, for ways to recompense reviewers for their service:

1   Publishing an annual list of reviewers who have reviewed for the journal over that year. This is becoming more common and is usually appreciated by the reviewers. For journals with large numbers of reviewers there will be little danger that reviewers can be associated with particular manuscripts and their anonymity compromised if closed peer review is used. However, in very small journals, with small reviewer databases, or where particular specialists are only used infrequently, care needs to be taken that the identity of reviewers is protected.

2   Sending Christmas cards or calendars to reviewers can have very positive effects (but care should be taken to make them and the greeting relevant and acceptable to various nationalities and cultures). Many reviewers are very grateful to receive them and may even write to thank the journal, saying that they have never before received something like this in all their reviewing experience. This is a terrible indictment of journals. It does take some effort and planning to produce, sign and mail out cards or calendars, but it means a lot to reviewers, and makes them feel that they are being thought of other than only when their help is needed with a review. It helps maintain a community spirit. An annual mailing can also be a visible recognition of their involvement with a journal if, for example, calendars are sent only to those individuals who have reviewed for the journal or have reviewed more than a certain number of manuscripts over the year.

3   Offering free offprints for the next paper the reviewer publishes in the journal.

4   Providing credit towards the publishing costs (such as colour-work or page charges) for papers the reviewer publishes in the journal.

5   Providing a gift (for example, a music compact disc or a book) after timely review of a certain number of manuscripts.

6   Inviting reviewers, or those who have reviewed a certain number of manuscripts, to an annual reception.

7   Giving reviewers a free life-long subscription to the journal after a certain number of manuscripts have been reviewed.

8   Offering reviewers for 'author-side-payment' journals a reduction in processing charges for their own manuscripts after they have provided timely reviews for a certain number of manuscripts.

How to develop and maintain reviewer loyalty

There are things that help keep reviewers loyal to a journal and make them more willing to review for it. Developing a reputation for quality and fairness is very important. There are also some other things that can help develop and maintain reviewer loyalty:

1   Always treat reviewers well and with courtesy. Remember that they are human. The following message from a reviewer illustrates that reviewers do notice and appreciate a courteous approach:

> Thank you for your email. I will be happy to review the paper. May I also thank you for the way you have requested this: I find it rather tiresome the way some editors these days email to inform me that I have been selected to review a paper or grant, that I am fortunate to be so selected and that it should be returned yesterday otherwise I have failed in my scientific duty – your politeness makes a pleasant change!

2   Ensure the reviewing workloads of your reviewers are monitored and that they are not overloaded or taken advantage of (see Chapter 4, page 52). Also remember that the manuscripts they receive from your journal might represent just a small proportion of the number they receive to review from all journals.

3   Respond to reviewers' queries promptly and sort out any problems they have as quickly as possible (see Chapter 4, page 70). Take their concerns seriously and never dismiss them out of hand or make them feel stupid.

4   Ensure the manuscripts reviewers are sent are always in good condition and complete, with all the materials they need included, and with clear instructions and guidance provided (see Chapter 4, page 60).

5   Ensure that manuscripts sent to reviewers have a reasonable standard of language (see Chapter 3, page 38) and are within the scope of the journal (see Chapter 3, page 36).

6   Provide feedback on manuscripts they have reviewed (see this chapter, page 139).

7   Give reviewers 'time' in return. Reviewers may contact a journal for assistance of various kinds and are very grateful when it is given. Examples are: providing details of the number of manuscripts they have reviewed, writing letters to official bodies in support of applications for citizenship to verify that they have reviewed for the journal, or sending them things they may not be able to get in their own countries and develop a craving for, such as certain chocolate bars! Requests must of course be reasonable, and legal, but as long as they are, take them all seriously and at least consider them.

Recognition of peer review as an accredited professional activity

There is growing interest in getting the importance of peer review more generally recognized. It is felt that it falls too far down the ladder of academic priority, and there have been calls for institutional support of peer review as an accredited professional activity denoting academic recognition and bringing credit to individuals as part of their *curriculum vitae* to carry weight in job applications, promotions, and funding. Some suggestions have been put forward as to how this could be achieved. One is that a system is needed whereby individuals can produce a peer-review record similar to their publication record. Weighting would be given according to the status or Impact Factor of the journals. This, together with review frequency, would be a measure of distinction. Institutions could produce a 'peer-review score', which would be taken into account in funding.[5,6] Another is that journals could send letters to their reviewers each year stating how many manuscripts they have reviewed, possibly with some measure of quality. This would be verifiable information that could be used as a criterion in assessment exercises.[7]

Some medical journals in the USA have started granting reviewers Continuing Medical Education (CME) credit for their reviewing work. CME is a scheme that follows American Medical Association guidelines whereby doctors must take part in continuing education to keep their licences. Credit for peer review was introduced in 2004 in recognition that reviewers feel they learn a great deal by doing it. For credit to be received, reviewers must demonstrate that they have been prompt and thorough in reviewing, and their review activities must be sponsored by an accredited provider working in collaboration with a medical journal indexed by the *Index Medicus*.[8]

So, the basic message is: treat your reviewers well. Journals could not survive without their expertise and the time and effort they put into every manuscript. This should not be forgotten.

References

1   Hames, I. (2001). Editorial Boards: realising their potential. *Learned Publishing*, **14**, 247–256.
2   Sense About Science. (2004). Peer Review and the Acceptance of New Scientific Ideas: Discussion paper from a Working Party on equipping the public with an understanding of peer review. Available for free download from www.senseaboutscience.org.
3   Schroter, S., Black, N., Evans, S., Carpenter, J., Godlee, F. and Smith, R. (2004). Effects of training on quality of peer review: randomised controlled trial. *BMJ*, **328**, 673. doi:10.1136/bmj.38023.700775.AE.
4   Training package for *BMJ* reviewers. http://bmj.com/advice/peer_review/ (accessed 30 May 2006).
5   Dominiczak, M. H. (2003). Funding should recognize the value of peer review. *Nature*, **421**, 111.
6   Clausen, T. and Nielsen, O. B. (2003). Reviewing should be shown in publication list. *Nature*, **421**, 689.
7   van Loon, A. J. (2003). Peer review: recognition via year-end statements. *Nature*, **423**, 116.
8   De Gregory, J. (2004). Medical journals start granting CME credit for peer review. *Science Editor*, **27**, 190–191.

8 The obligations and responsibilities of the people involved in peer review

Various parties are involved in the peer-review process: authors, editors, reviewers, and editorial office staff. All have certain obligations and responsibilities, and should always act according to the highest ethical standards (see Golden Rule 9). No one should use any information they receive during the submission and peer-review process for their own or others' advantage or to disadvantage or discredit others (see Golden Rule 10). All should also declare any potential conflicts of interest and excuse themselves from involvement with any manuscript they feel they would not be able to handle or review objectively or fairly (see Golden Rule 11). Some of the obligations and responsibilities are common across all the groups, others apply to just one or other. Authors, editors and reviewers may feel they know what is required of them in the various roles – many will probably be acting in more than one of these at any one time or at different times of their lives – but they may be hard pressed to list any other than the most obvious if asked. Many societies and professional bodies publish codes of conduct or ethical guidelines for their members. Some are general, but most are aimed at the specific activities relevant to those organizations. Quite a number do, however, include guidance on research and publication issues and so are

Golden Rule 9

Everyone involved in the peer-review process must always act according to the highest ethical standards.

Golden Rule 10

Information received during the submission and peer-review process must not be used by anyone involved for their own or others' advantage or to disadvantage or discredit others.

Golden Rule 11

All the parties in the peer-review process must declare any potential conflicts of interest and excuse themselves from involvement with any manuscript they feel they would not be able to handle or review objectively or fairly.

well worth looking at: for example, those of the American Chemical Society (www.chemistry.org), American Geophysical Union (www.agu.org), American Mathematical Society (www.ams.org), American Physical Society (www.aps.org), and the Institute of Electrical and Electronics Engineers (www.ieee.org). Excellent and extensive guidelines, with specific examples, can be found on the websites of both the Council of Science Editors (www.CouncilScienceEditors.org) and the International Committee of Medical Journal Editors (www.icmje.org). It will be helpful here to go over the general obligations, responsibilities and ethical standards that the various parties involved in peer review should consider and follow.

Authors – their obligations and responsibilities

Although this book is not aimed at the people submitting manuscripts to journals, i.e. authors, these individuals may contact editors and editorial office staff for advice on various issues connected with their general obligations and responsibilities in the authorship role. It is important, therefore, for editors and editorial staff to know what these obligations and responsibilities are. Authors can also be directed to the guidelines on good research practice published by a number of organizations, for example by the Wellcome Trust.[1]

For working scientists, research will be a major component of their lives and some projects may have taken many years; they will have invested much time, energy and financial resources in these. The time will come when they feel a piece of work is reaching the point where it is complete enough to publish, and they will think about writing it up in manuscript form to submit to a journal. What are their obligations and responsibilities?

To act honestly

Authors should be submitting original work that has been honestly carried out according to rigorous scientific standards. It should not have been obtained fraudulently or dishonestly, or have been fabricated or falsified (see Chapter 9, page 174). When they write up their work for publication, authors should present a concise and accurate account of how the research was carried out, the results that were obtained and an objective discussion of their significance. There should be enough detail to enable the work to be repeated by others, data should be accurately reported, never 'fudged', and problematic data should not be left out selectively so as to provide a 'clearer' story. Originality should not be claimed if others have already reported similar work or aspects of it. Credit should always be given to the work and findings of others that led to the work or influenced it in some way. Authors should not present their work, or use language, in a way that detracts from the work or ideas of others. Information obtained privately should not be used without the explicit permission of the individuals from whom it was obtained, and appropriate letters confirming permission to include this information must be acquired for journals that require this. Authors should declare any conflicting interest when they submit their manuscripts, whether or not required to do so by the journal (see Golden Rule 11 and this chapter, page 164).

To choose the most appropriate journal

Authors should choose the most appropriate journal to which to submit their work. This may seem so obvious that there is no need to state it, but it is a very important point, and one that some authors regularly fail to observe. The journal an author chooses should primarily be one for which the subject area is suitable. The scope of journals can change over time, so some investigation may be needed to find out current policy. Guidelines for authors should be read, and recent issues checked for the sorts and quality of papers being published. Authors should never be frightened of contacting editors to ask whether or not a potential submission would be suitable for their journals. Many journals actually have procedures in place for such pre-submission enquiries (see Chapter 3, page 26). The novelty and significance of the work should be appropriate for the journal – authors should avoid making exaggerated claims about the novelty or significance of their findings in the hope of getting it past any initial appraisal, or even past the reviewers. Specialist reviewers will be wise to this sort of tactic and will get impatient with authors who regularly make inflated claims about their work.

Young scientists at the start of their careers should be guided by their mentors, but it is to be expected that there may initially be some trial-and-error element to their submission behaviour – this is all part of the learning process and many supervisors are willing to go along with a journal choice that is not totally appropriate if pressed because of this. Experienced and established researchers should not, however, always be trying to submit to journals whose standards for acceptance are way above what they are submitting – editors will soon pick up on this and they and the reviewers will very soon become irritated with this 'give it a go' type of approach.

To make sure manuscripts are well presented, contain nothing inappropriate and are submitted correctly

Once a journal has been chosen, authors should familiarize themselves with the specific submission requirements of that journal. Manuscripts should be submitted in the format requested, to the appropriate place and in the correct way, with all the required information provided and all necessary enclosures included. If anything has not been provided or the manuscript deviates in any way from the stipulated requirements, this should be noted and explained in a cover letter. For example, if there are length restrictions and a manuscript exceeds these, authors should explain why and present justification. This will allow the editor to evaluate whether or not there is a valid case for leniency and also to ask the reviewers, or give them guidance, about this aspect if necessary. Manuscripts should be checked carefully before submission for language and correct presentation – editors and reviewers should not receive sloppily prepared manuscripts full of errors that could have easily been picked up on a final careful read through. Authors are also responsible for ensuring that their manuscripts do not contain plagiarized material or anything that is libellous, defamatory, indecent, obscene or otherwise unlawful, and that they do not infringe the rights of others. Authors have a duty to check the references they cite very carefully to ensure that the details are accurate. They should not take these from the reference lists of other authors and so perpetuate errors that previous authors may have introduced into the literature. If investigations have involved animals or human subjects, authors should provide all the statements required by journals that the experimental protocols were approved appropriately and meet the guidelines of the agency involved, and informed consent was obtained where required. It is the responsibility of the authors to check a journal's requirements and to obtain and provide confirmation of compliance with all policy issues relating to publication in that journal. Any restrictions or failure to provide certain items or non-compliance with journal policy should be mentioned at the time of submission. Submission to a journal by authors is taken to indicate that they agree to abide by the regulations, policy issues and publishing requirements of that journal, even if this is not explicitly stated.

It is inevitable that all authors will have some of their manuscripts rejected. They may choose to carry out the revisions required if a journal has indicated that the paper would be or may become acceptable with these revisions and then resubmit the manuscript to that same journal. They may, however, decide to submit to another journal, especially if timing is crucial and they need to get the work published as soon as possible, or if they are no longer in a position to be able to carry out any more experiments if additional data have been called for as part of the revisions required. Authors should then reformat the manuscript to meet the requirements of the new journal, and the cover letter should be redrafted. Every editor and editorial staff member will be able to relate cases where manuscripts have arrived with sections in the wrong order, references in the wrong format, and cover letters still bearing the name of another journal or journal editor. This is highly discourteous and immediately gives the impression that the authors can't be bothered to make any effort but are expecting the journal and its editors and reviewers to invest time and effort in their manuscript.

A more serious problem arises when authors submit to another journal without taking on board the reviewers' comments from a previous submission elsewhere. Some criticisms may have been serious and not just a case of the manuscript having insufficient novelty or general interest for the journal's audience. For example, serious deficiencies or flaws may have been picked up, ones not resolvable without the addition of extra work or a re-evaluation of experimental design. Others may be trivial and should be corrected. It may be that a manuscript will be sent out by a subsequent journal to a reviewer who has already seen the manuscript – reviewers tend, with good reason, to get very annoyed when the comments in their review of a manuscript, which will have involved considerable time and effort, have not been addressed. All presentational errors, such as incorrect citations, inconsistency in dates, and poorly phrased or ambiguous sentences, should always be corrected before submission elsewhere – there is absolutely no excuse for not attending to these before resubmitting.

Authors should not submit the same work or paper to more than one journal at a time (dual or multiple submission; see Chapter 9, page 179). This may seem obvious to many, but it is a question I've been asked by young researchers so it is something about which supervisors need to educate their students. Authors should also not submit a paper that overlaps substantially with ones already published (so-called re-dundant or duplicate publication; see Chapter 9, page 178). Work already published should be referred to, and the full citation(s) given. If authors have related manuscripts submitted or in press elsewhere, they should mention this, giving full details. Many journals require that copies of such manuscripts be included with the submission to them, and authors should be sure to include them at the time of sub-mission (see Chapter 3, page 30). If they are not sent to the reviewers with the manuscript under review, delays can arise because the reviewers may request them or else return their reviews saying they were unable to provide an accurate assess-ment of the work because they did not have access to the related manuscript(s). Authors should also not try to divide up papers inappropriately into smaller ones (minimum publishable units or MPUs) in an attempt to increase their publication records. This so-called 'salami publishing' leads to inadequate papers that will irritate editors and reviewers, and very often lead to rejection. It can also lead to lost oppor-tunities for authors to publish their work fully, as a complete story, in higher-ranking journals.

To deal appropriately with all authorship issues

Establishment of authorship and responsibility
Authorship and issues related to it are very important and can become thorny prob-lems. Authors may seek advice from a journal before submitting their manuscripts. As a general rule, all individuals named as authors should qualify for authorship and all those who do qualify should be listed. But what qualifies a person for authorship? Many journals recommend that authors follow the Vancouver Guidelines on author-ship, as defined in the International Committee of Medical Journal Editors' (ICMJE) Uniform Requirements for Manuscripts Submitted to Biomedical Journals.[2] These state that authorship credit should be based on:

1) substantial contributions to conception and design, or acquisition of data, or analysis and interpretation of data;
2) drafting the article or revising it critically for important intellectual content; and
3) final approval of the version to be published.

The Guidelines stipulate that all these criteria should be met. Getting funding, collection of data, or general supervision of the research group do not, on their own, qualify a person for authorship. They also advise that all contributors who do not meet the criteria for authorship should be listed in the Acknowledgements section.

Until recently, all authors were considered to be responsible for all the contents of a paper. However, in response to some high-profile misconduct cases in physics where several co-authors were cleared of any misconduct,[3] there has been a move for authorship guidelines to be changed to reflect that the responsibility for the integrity of a scientific paper does not always need to be carried by all the authors. The ICMJE Uniform Requirements encourage editors to develop policies for identifying who is responsible for the integrity of a work as a whole and stipulate that 'Each author should have participated sufficiently in the work to take public responsibility for appropriate portions of the content'. In November 2002, the American Physical Society (APS) adopted new guidelines and these (which still stand in 2006) may be suitable for many journals. Basically, they recommend that all co-authors share some degree of responsibility for their paper, but only some need to have responsibility for the whole paper. The following excerpt is reprinted with permission from the American Physical Society (APS Ethics and Values Statements: 02.2 APS Guidelines for Professional Conduct[4]. Copyright 2006 the American Physical Society):

> All collaborators share some degree of responsibility for any paper they coauthor. Some coauthors have responsibility for the entire paper as an accurate, verifiable, report of the research. These include, for example, coauthors who are accountable for the integrity of the critical data reported in the paper, carry out the analysis, write the manuscript, present major findings at conferences, or provide scientific leadership for junior colleagues.
>
> Coauthors who make specific, limited, contributions to a paper are responsible for them, but may have only limited responsibility for other results. While not all coauthors may be familiar with all aspects of the research presented in their paper, all collaborations should have in place an appropriate process for reviewing and ensuring the accuracy and validity of the reported results, and all coauthors should be aware of this process.
>
> Every coauthor should have the opportunity to review the manuscript before its submission. All coauthors have an obligation to provide prompt retractions or correction of errors in published works. Any individual unwilling to accept appropriate responsibility for a paper should not be a coauthor.

Submitting authors have a responsibility to obtain agreement from all co-authors on the final version of the manuscript and on its submission. Some journals require written confirmation of this.

It has been suggested that there should be a move away from a prescriptive approach to a more descriptive approach to authorship. A number of biomedical journals have moved to the concept of 'contributors' rather than authors, with details being given of the exact contribution of each to the planning, conducting and reporting of the work. It is felt that this exact attribution is a much fairer system and will encourage honesty and dissuade fraud. It will also help in academic appointments and promotion as it makes it possible to assess accurately someone's exact contribution to a paper. (The excellent article by Rennie *et al.* covers many authorship issues and presents the case for making authors more accountable.[5]) Authors should keep accurate records of who did what throughout a research project, as this will make it easier to assign attribution when the time comes to submit the work to a journal. It will also make it easier to resolve author disputes.

The ICMJE Uniform Requirements now strongly encourage editors to develop and implement a contributorship policy, but they recognize that this 'leaves unresolved the question of the quantity and quality of contribution that qualify for authorship'. The *BMJ* (*British Medical Journal*) lists contributors in two ways (see http://bmj.bmjjournals.com/advice/article-submission.shtml#author): it publishes a list of authors' names at the beginning of the paper and then lists contributors, some of whom may not be included as authors, at the end of the paper, giving details of who did what. Other medical journals, for example *The Lancet*, also list the contributions of the authors listed and other contributors. Whether, and how widely, this system will be adopted by other disciplines remains to be seen. The *BMJ* also asks that contributors on a paper nominate one or more 'guarantors', who will have overall responsibility for all the work. Guarantors, as described in Rennie *et al.*,[5] are individuals who have contributed substantially to the work but who also ensure integrity of the whole project. They must be accountable and take responsibility for all parts of the manuscript, both before and after publication. The authorship policy at some other journals, for example the multidisciplinary science journal *Nature*, is that authors are strongly encouraged to include a statement to specify the actual contribution of each co-author. However, only partial adoption has been seen in these voluntary schemes. In the wake of increasing numbers of cases of serious fraud, the top science journals are looking at making detailing contributorship an obligatory requirement of submission.

Various other schemes for authorship emerge from time to time. For example, a quantitative method for evaluating authorship has been suggested (QUAD, Quantitative Uniform Authorship Declaration), based on four categories of contribution: conception and design, data collection, data analysis and conclusions, and manuscript preparation.[6] In this scheme, each author is attributed with a percentage share of the total credit in each of these categories, and authors are listed in descending order of total contribution across all four categories. The least an author can contribute is 10% within a single category, placing a theoretical limit of 40 on the total number of authors. The proposers of this system argue that its quantitative nature and transparency should help reduce abuses of authorship listings.

Each group of authors needs to decide who is to be the person who will act as the corresponding author during submission and review. A journal cannot correspond individually with each author (see Chapter 5, page 102), and it is up to the corresponding author to ensure that all authors who should be are included, the order has been agreed, and all the co-authors are happy with this. The authors should have resolved any disagreements or problems before submission. Journals should not become involved in authorship disputes. Many journals have check boxes for authors to fill in for statements relating to authorship. Online submission systems can be set up so as to halt submission until various statements have been confirmed. It's a good idea to list the authorship statements here along with others about compliance with journal policy (see Chapter 6, page 125, and Box 6.2): for example, that the work being presented is original and not published, in press or submitted elsewhere; that the manuscript has been prepared according to journal instructions and policy; and any on specific areas such as agreement to release upon request data and/or biological and chemical materials that are in the paper but not available commercially to others for non-profit research, and declarations that materials have been deposited in recognized repositories where appropriate or that trial numbers have been obtained for randomized controlled trials.

Unacceptable 'authorship'

Readers of journal articles should feel confident that the people who are listed as authors had legitimate input into the work and manuscript and that everyone who was involved with the work and preparation of the manuscript is listed. There are two cases when this is not so – 'gift' and 'ghost' authorship – and these types of 'authorship' are not acceptable.

In 'gift authorship' (also known as honorary or guest authorship), someone's name is added to the author listing even though they may have contributed very little, or perhaps nothing, to the research or the writing of the manuscript. They do not meet recognized authorship criteria. This applies even if they were responsible for obtaining the funding if that was all they did. Why do people give away authorship? If they do it willingly, for example by bestowing gift authorship on a senior researcher or the head of their department or institute, they may feel this gives greater authority and prestige to their paper. Or they may be doing it to keep in favour with that person. They may, however, be doing it unwillingly, under pressure from certain individuals or to comply with unwritten local 'rules'. Until recently, for example, especially in some countries, it was not uncommon for senior researchers, group leaders or institute heads to be automatically included as authors on many papers. However, the occurrence of various high-profile fraud cases in a number of disciplines and countries may have put a brake on this. Some senior authors have claimed immunity in these fraud cases by virtue of not knowing what was in the papers! Surely a lesson for all authors to make sure they know and approve of what is in any paper carrying their name? Companies may sometimes invite a well-known or prestigious researcher who has not had anything to do with some work or the manuscript to submit and publish it under his or her own name to give it extra credibility and impact (i.e. giving that person guest authorship). This, with its total lack of transparency and accountability, is unethical practice.

'Ghost authorship' occurs when a person who has made a substantive contribution to the research reported in a paper or to its conception and the writing of the manuscript is not named as an author. There are two types of ghost authorship. In the first case, someone may be kept off the author listing without their knowledge or against their will. This is not fair and can be considered to be misconduct by the person or people keeping someone's name off the list if that person merits being named as an author. In the second case, someone knows and agrees that they will not be named as an author even though their contribution warrants inclusion. They may be doing it in return for payment or some other form of recompense at someone else's request – for example a company, which then goes on to invite someone else to be the author (see 'gift authorship' above, page 154).

Order of authors
There is evidence that there is an ever-increasing rise in the number of authors per article in science and medicine, and a substantial decrease in the number of single-author papers. Weller gives details from various studies on the change in number of authors over time since the 1930s.[7] In medicine, for example, the number of authors per article (averaged from 14 studies) was 1.4 in the 1930s and 4.6 in the 1990s. Cho and McKee report that for biomedical research papers, the average number of authors in 1930 was 1.3 and this had risen to 6.0 by 1989.[8] Nowadays, as a consequence of collaborations across countries and disciplines and costly and ambitious projects in 'big' science, the numbers of authors can run into the hundreds, even thousands. The position an author's name occupies is therefore very significant. However, there are no set rules on how this should be decided and conventions vary between disciplines. In general, and certainly in the life sciences, the person who has done most of the work, usually a graduate student or post-doctoral fellow, will be first and the mentor of the project in whose laboratory the work was carried out will be the last. But this is very much a generalization and the variations are numerous, especially between disciplines and with multi-group/centre submissions. With the latter, there may also be a group name, which will be indexed along with the names of the individuals in that group. For example, in the paper reporting the complete human genome sequence, the author byline is given as 'International Human Genome Sequencing Consortium', with a note indicating that the list of authors and their affiliations are in supplemental information online.[9] The 18 centres that participated are given, with a listing of authors (who run into the thousands) beneath each. In other group work, names may be listed according to the roles taken in the project rather than by institution, with the affiliations listed separately. This is done, for example, in the paper reporting the draft chicken genome sequence.[10] The byline is 'International Chicken Genome Sequencing Consortium' and the categories of involvement are: overall co-ordination; genome fingerprint map, sequence and assembly; mapping; cDNA sequencing; other sequencing and libraries; analysis and annotation; and project management.

It is up to each individual group of authors to resolve all authorship issues, and all the authors need to be satisfied with the final listing before submission of the manuscript. Disputes on authorship can be very serious and have sometimes ended up in court, such as the case where one author substituted her name for another's as first author on the final draft of a manuscript without approval from the co-authors.

The original first author took legal action and submission of the manuscript was prevented by court injunction until the court ruled.[11] Such delays in publication can have seriously detrimental effects on the work and careers of the co-authors involved. In an attempt to get round the problem of authorship, some editors have tried a different system and listed authors alphabetically. Unfortunately, it's been found that authors whose names fall later in the alphabet may avoid sending their papers to such journals![12]

Changes in authorship after submission

Occasionally a journal may receive a request for a change or changes in authorship after the manuscript has been submitted. This may come from the corresponding author or from one of the co-authors, or even from someone who is not on the author listing. Journals should obtain reasons for any requests for changes in authorship and ensure changes are legitimate and justified. Some journals, for example those of the American Physiological Society (www.the-aps.org), actually require all authors to fill in and sign their agreement on a 'Change of Authorship Form' (see Appendix II, page 220). Requests for additions from one of the authors are the most common and are generally straightforward – often just a case of genuine omission – but they must be authorized by the corresponding author. Requests from non-authors and for removals, however, need to be looked at carefully and appropriate action taken. Possible cases include:

1   The corresponding author may request a removal. The reason for this needs to be obtained, if not provided, and evaluated, but it is often simple and reasonable. Frequently, part of the work or results that had been included in an earlier version of the manuscript is no longer there, but the name of the person who carried out the work wasn't removed before actual submission. This can also very easily happen during the revision of a manuscript. Confirmation should be obtained from the author being removed that this is being done with their knowledge and agreement.

2   A co-author may ask for their name to be removed. Again, reasons need to be given. These may range from the trivial to the serious, and may have implications for the review of the manuscript. For instance, the author may say that the manuscript was submitted without their approval because they felt the data were inadequate, problematical or did not support the interpretations and conclusions made. The editor will need to assess the situation and decide whether or not to notify the reviewers of this or to wait for their reviews to come in and then evaluate them in the light of the new information. The editor should also contact the corresponding author for a response to the co-author's request and the reasons given for it.

3   A co-author may complain that his or her name is not in the correct place in the author listing and request that it be moved, for example to first-author position. This should be referred to the corresponding author, who is responsible for any disagreements between the authors on authorship and for resolving disputes.

4   Someone who is not even on the list of authors may contact a journal to say that if paper X from group Y is under review, their name should be on it. This is the most difficult of all situations. Firstly, the person making such a request cannot be given any information on whether any such manuscript has even been submitted to that journal – that is privileged and confidential information (see Golden Rule 3). If the manuscript has been submitted to the journal, the editor will need to raise the matter with the corresponding author and ask for a response. Such issues need to be dealt with on a case-by-case basis, with the editor acting to ensure integrity and correctness of the submission. Serious allegations that cannot be answered adequately by the submitting authors may lead an editor to withdraw a manuscript from review until the inter-researcher issues have been resolved.

Pre-submission and post-acceptance considerations
There is concern in certain disciplines that as links between industry and researchers increase, scientists will be compromised as to what they can or cannot publish. The ICMJE Uniform Requirements for Manuscripts Submitted to Biomedical Journals state that 'researchers should not enter into agreements that interfere with their access to the data and their ability to analyze it independently, to prepare manuscripts, and to publish them'.[2] In the medical field, there are concerns about bias being introduced by, for example, pharmaceutical companies insisting on publishing only positive and not negative data. Some journals may therefore require a statement from authors declaring that they have not been subject to this kind of compromise or been influenced by any sponsor of the research in their decisions on what to publish, and when to publish. It is the researcher, not the sponsoring company, who holds control of the data. The ICMJE Uniform Requirements also recommend that authors 'describe the role of the study sponsor(s), if any, in study design; in the collection, analysis, and interpretation of data; in the writing of the report; and in the decision to submit the report for publication'. If the sponsor had no such involvement, this should be stated.

Once an author's manuscript has been accepted, no significant changes should be made without the approval of the editor or journal editorial office, whose staff should refer to the editor anything other than minor textual changes (see Chapter 5, page 111). This could happen, for example, when the final copy for publication needs to be provided in separate electronic form after acceptance of a paper or online submission. The final copy for press should correspond to that accepted. If any changes are made, they should be outlined in a letter. Notes added at proof stage should also be checked with the editor (see Chapter 5, page 111).

Authors should abide by any policy the journal has on the free distribution of materials and/or data reported in the paper (see Chapter 3, page 30). These were the conditions under which the paper was submitted, reviewed, and accepted, and the authors are duty bound to adhere to them. The free availability of material and data is considered by many journals to be essential for the advancement of science. Occasionally, people may contact a journal complaining that they have been unable to obtain materials from authors despite repeated requests. It is up to each journal to decide how to deal with this. Often, a letter from the journal is sufficient to elicit the materials. However, editors and journal staff should be wary, as the case presented by the person requesting and not receiving materials may not be an accurate account of the situation. It may even be that the requestor is not getting any response to letters because the author's contact details have changed. This may particularly be so when communication is by email, as email accounts change. A recent study suggests that one in four email addresses of corresponding authors becomes invalid within a year of publication.[13] There may also be resource considerations (see Chapter 9, page 181). Diplomacy and tact are often called for to avoid alienating either party or upsetting authors with accusations of non-compliance with journal policy that may prove to be unfounded.

After publication, authors have a responsibility to answer questions about their work and paper and to resolve any issues that arise. If errors are found, it is the author's responsibility to ensure the journal is notified immediately so that a correction, or a retraction in extreme cases, can be published as soon as possible if necessary (see Chapter 9, page 192). With paper journals it is always difficult to ensure that someone reading an article will get to see the correction. But with the advent of online journals, links can be set up directly to correction notes, which will therefore be read at the same time as the article.

Editors – their obligations and responsibilities

General responsibilities

Editors-in-chief (for ease and clarity, the term 'Editor' will be used to denote editor-in-chief in this section) are primarily responsible for ensuring the quality of their journals and that what is reported is ethical, accurate and relevant to their readership (see Golden Rule 1). They are the 'gatekeepers' of their journals, and are responsible for everything to do with them. As they have the ultimate decision-making powers, their behaviour must be transparent and totally beyond reproach. They must not abuse the trust of any of the parties involved in the peer-review process. They must not use privileged information for personal gain or to discredit others. Editors need to ensure that they have total editorial independence (see Golden Rule 7) and are not influenced by any pressure group. They (and all editors) should also judge each manuscript submitted to them solely on its merits, without regard to the race, gender, religious belief, ethnic origin, citizenship or political philosophy of the authors (see Chapter 5, page 90, and Golden Rule 8). When they submit a manuscript to their own journal, they should remove themselves completely from its review and not attempt to find out details of the review process or to influence review or the decision.

Editors are responsible for developing editorial policy, usually in consultation with their editors, and for amending and updating this to take account of changes in their field and in scientific publishing in general. They need to formulate editorial policy so that all the editors are working consistently and so that quality is maintained (see Chapter 3, page 26). They should also aim to attract the best and most appropriate people to their editorial teams. No Editor should ever attempt to increase the Impact Factor of his or her journal by unethical means, for example by asking authors either to include more citations to the journal irrespective of whether they are warranted or to delete those to competing journals, and they should never imply that acceptance will depend on compliance with such requests. This is all highly unethical and should be considered misconduct (see Chapter 9, page 183).

Editors have a duty to ensure that papers submitted comply with recognized ethical guidelines, and that all procedures at their journals are ethical and in accordance with best-practice recommendations. They need to monitor all aspects of their own journal's activity and deal with any problems that arise to ensure that manuscripts are handled and published in a timely manner. They should put in place, if they do not exist, procedures for dealing with suspected misconduct or fraud, including what action will be taken if misconduct has been shown to occur (see Chapter 9, pages 184 and 190). Editors are nowadays frequently called upon to provide guidance on many aspects of scientific publishing and so should keep up to date on issues relevant to ensuring that the quality of their publications is maintained and that fair procedures are employed.

Responsibilities to authors

Authors entrust journals, and so all editors, with the results of their work, and therefore indirectly with their reputations and career prospects. This trust must not be abused. All manuscript submissions and their contents should be kept confidential (see Golden Rule 3) and the Editor should ensure that everyone involved in the handling and review of manuscripts understands this and that they are dealing with privileged information that must not be used for private purpose or gain.

Authors should be treated fairly and courteously, and with objectivity – if there are any personal or professional conflicts, editors should not be responsible for the review of and decision making on those manuscripts, but should hand them over to another editor to handle. This applies in cases of potential positive as well as negative bias. Authors who are friends, close collaborators, or from the same departments or institutions should not be exposed to the possible criticism that their papers have received unfair positive treatment because of any relationship with the editor. This can be harmful to authors, and undermine the confidence with which their work is viewed. The review of their manuscripts and the resulting decision-making process must always be able to withstand rigorous scrutiny.

All editors have a primary responsibility to authors to ensure the efficient, timely and fair review of manuscripts submitted to them. They should choose the most appropriate and expert people to act as reviewers. Authors' requests that certain people be excluded from review should be considered, and the reasons for such requests borne in mind if these people are used in the review process. Editors must never deliberately choose reviewers who they know will provide either a favourable or an unfavourable review. Similarly, notoriously slow reviewers should not be chosen so as to deliberately hold up the review of any manuscript. To protect authors' interests, individuals who are consistently slow in review or who regularly fail to return reviews should be removed from a journal's active reviewer database unless there is a very good reason not to, for example if there are very few experts in a certain area or those individuals have a unique combination of talents or expertise (in which case they are probably inundated with requests to review from other journals).

Editors should ensure that all editorial policy and instructions for submission are clearly set out so authors can gain easy access to them before they submit their manuscripts. Authors should be aware of a journal's policy on peer review and which system is in use (closed or open; see Chapter 4, page 41). They should also be aware of what is acceptable in terms of preliminary reports elsewhere or the posting of material on websites, and what would exclude any submission being considered for publication. This will vary from journal to journal. It is also an area that is under considerable change, so journals do need to make their policies very clear (see Chapter 3, page 26, and Chapter 9, page 178).

All authors' queries should be dealt with promptly and courteously, and explanations or clarification provided when requested on procedures or on the decision, or when any unexpected delays occur during review (the reasons for these should be provided). In cases of accusations of misconduct, authors should be notified of these and given the opportunity to respond and put forward their side of the story before any punitive action, such as notifying their departments or institutions, is taken (see Chapter 9, pages 184 and 190).

Decisions should be communicated in a timely and efficient manner and all conditions for acceptance subject to revisions should be made clear. If there are any reasons or conditions that would prevent eventual publication, authors should be made aware of these so that they can choose to submit their manuscripts elsewhere if they know they will not be able to meet the required criteria.

Responsibilities to reviewers

Editors (and journals) should ensure that reviewers are sent manuscripts that are appropriate to their area of interest and expertise. They should be provided with clear instructions and guidance on the journal's aims and scope, and on what is expected of them in the review process. They should also be briefed on any specific areas for which their assessment is being sought, and be provided with all the information and ancillary material that they may need to provide a thorough and efficient review of a manuscript. Reviewers should be made aware of anything relevant that comes to light during the review process. If a manuscript is withdrawn from review, by either the authors or the journal, they should be notified immediately to avoid unnecessary extra work on their part. Reviewers should be made aware of the time normally allowed for review, and if a longer timeframe is agreed they should not be bombarded with inappropriate reminders to return their review. If reviewing forms are used, these should be well thought out and developed so as to help the reviewer submit a thorough and meaningful review.

If a journal operates open review (see Chapter 4, page 42), reviewers should be made aware of this and it should be ensured that they understand exactly what this means, namely that their names will be made known to the authors, before they agree to review. If open peer review is not used, reviewers' anonymity must under all circumstances be protected and not compromised (see Golden Rule 4). Editors should alert reviewers to any possible means by which their identity could be compromised, for example by accessing authors' websites to view data or tools posted there (see Chapter 4, page 64). In such cases, Editors must find alternative means to provide reviewers with the data and other materials they need to provide a thorough and accurate assessment of the manuscript. They should also not pass on electronic files from the reviewers to the authors through which their identity could be determined (see Chapter 4, page 74).

Editors should be aware of reviewers' workloads, both current and past, so that no single reviewer is asked to carry too heavy a workload (see Chapter 4, page 52). Reviewers should feel reassured that someone is monitoring this and looking after their well-being and not taking advantage of them. Reviewers should not be made to feel that they can never say 'no' when asked to review, or that this will have repercussions for their own manuscripts.

Editors should try to find ways to thank reviewers and recognize their contributions, both privately and publicly. At individual manuscript level, they should be thanked and, if possible, informed of the outcome of the review process (see Chapter 7, page 139). On a general level, any requests reviewers make of a journal should be considered and complied with if they are reasonable and it is in the journal's power to do so, for example providing letters listing their service to the journal that reviewers may need to further their careers or to help them in applications for naturalization or residency. Ways of thanking and rewarding reviewers are covered in Chapter 7 (page 142).

Responsibilities to readers

Readers should be able to assume that what they read in a peer-reviewed journal is accurate and is a valid and honest contribution to the literature. It is the Editor's responsibility to, in effect, provide the guarantee for this. There should be enough evidence and detail presented to allow readers to evaluate the authors' conclusions and to repeat the experiments. This is how scientific progress is made and how a field moves forward; building on previous work, correcting errors or highlighting inconsistencies. Readers should have confidence that articles have been rigorously and fairly peer reviewed, and that corners have not been cut in an attempt to get 'hot' papers submitted or to beat a competing journal into print. If an Editor is not an expert in the subject area of any manuscript, and they cannot nowadays be expected to be experts in all the areas covered by their journals, it is their responsibility to seek appropriate advice, for example from a section editor or external experts.

Editors have a responsibility to their readers to make sure that the reviewers they use are of high standard; they should stop using any who fail to return reviews or routinely provide late, superficial or inadequate reviews. They should reprimand and penalize those they find breaching peer-review ethics, for example reviewers who have disclosed confidential information contained in manuscripts previously sent to them or used it to their own advantage (see Chapter 9, page 192). Editors have a responsibility to monitor whether there are any conflicts of interest of reviewers, or of anyone else involved with a manuscript during its review; to have full information from authors about any potential conflicts they may have; and to obtain reassurances on independence of action and control if necessary. If a case of alleged or suspected misconduct or fraud is brought to the attention of the Editor, he or she has a duty to look into this and to raise it with the author and obtain a response (see Chapter 9, page 184). It is not acceptable for Editors to take no action and ignore suspicions or accusations. If any suspicions are found to be valid, the Editor should make sure that appropriate corrective measures are put into place. Readers should be informed of any work that is found to be fraudulent, fabricated, plagiarized, or the result of serious misconduct (see Chapter 9, page 192).

Editors should ensure that readers are able to distinguish between peer-reviewed and non-peer-reviewed material in their journals. They should make clear any sponsorship of articles.

Editors have an enduring responsibility to ensure that authors abide by all the policies of their journals that are of relevance to readers, such as, for example, the provision of research materials reported in papers (see this chapter, page 157). If there is non-compliance with the conditions set out by the journal, and which authors agreed to by virtue of submitting to the journal, Editors have a duty to pursue this with the authors and to implement appropriate sanctions if non-compliance persists.

Reviewers – their obligations and responsibilities

The fate of any manuscript will depend largely on the reviewers' assessment and what comments and recommendations they return. They are therefore of central importance in the peer-review process and as such need to be trustworthy and honest in how they treat and assess manuscripts. When first approached to review a manuscript they should declare any conflict of interest, either real or perceived (see Golden Rule 11), to allow the editor to decide whether this should disqualify them from review. Reviewers should disqualify themselves if they feel unable, for any reason, to provide an honest and unbiased assessment. They should not feel bad about declaring such a conflict of interest – editors recognize that everyone is human and they will appreciate reviewers' honesty.

Reviewers have a responsibility to let editors or the appropriate people at a journal know if a manuscript falls outside their area of expertise, or if they are able to review just specific parts of it. They should also let the editor or journal know if they will not be able to review within the requested timeframe so that other potential reviewers can be approached if need be. If they have agreed to review a manuscript, they should alert the editor or the journal if any unexpected circumstances arise that will delay their review. They should not rush and submit an inadequate or superficial review just in order to meet their review deadline.

Reviewers should read all the accompanying material, instructions and guidelines sent with a manuscript before doing the review, and they should contact the journal if anything is not clear or if any items are missing. If they feel they would like to seek the advice of a co-worker or colleague, they should always first check with the journal to ask for permission to do this – there may be a reason why that person should not be involved in the review of that manuscript, or even know of its submission. If a colleague does share the reviewing workload, it is only fair that the co-reviewer receives credit for this – the journal can add their name to the reviewer database and include them in any 'thank you' or remuneration procedures in place (see Chapter 7, page 142).

Reviewers must keep manuscripts and accompanying material confidential (see Golden Rule 3); they should not leave copies lying around for others to see or pass copies to colleagues or members of their groups. They should not use any of the information for their own or others' advantage or to disadvantage others (see Golden Rule 10), and they should not plagiarize any of the material. They should destroy the paper copy or delete the digital file of the manuscript after they have finished their review.

Reviewers should be objective and constructive in their reports and refrain from making personal comments or defamatory remarks (see Golden Rule 6). They should explain their judgements and support them with evidence so that they can be easily and accurately evaluated by the editor, and so that they will be helpful and constructive to the authors. They should let editors know if a manuscript they receive for review overlaps significantly with or is very similar to one that has already been published. They should report to the journal instances of suspected plagiarism, fraud, or other misconduct, and ask for advice on how to proceed. Reviewers should not, except for legitimate reasons, request that authors include citations to their own work in order to receive additional citations for themselves.

And finally, all scientists should remember that they have an obligation to review their fair share of manuscripts, for when they are authors other scientists will be acting as reviewers for their manuscripts. It is very much a give-and-take situation. Reviewers nobly give their valuable time, usually without any form of conventional 'payment', to assess the work of others and to provide valuable feedback and suggestions for improvement, which they in return receive when they submit papers for publication. The peer-review process and publication of authenticated scientific work could not continue without this review by real experts.

Editorial office staff – their obligations and responsibilities

The staff in editorial offices have an obligation to treat all submissions and anything to do with their review in confidence (see Golden Rule 3) and to act with discretion at all times. They should always behave professionally, and treat all authors, reviewers and editors with courtesy and respect. They should abide by the editorial policies of their journals, apply those policies consistently, and act according to best-practice guidelines in all areas of activity. They should help their Editors achieve their vision for the journal and act in the best interests of the journal if any conflict arises. As with all other parties involved in peer review, they should declare any potential conflict of interest and excuse themselves from dealing with any manuscript they feel they would not be able to handle objectively or fairly (see Golden Rule 11).

Conflicts of interest – what they are and how to deal with them

Because of the enormous impact publication can have, not only on the reputations and careers of individuals, but also on political decision making, policy formulation and financial returns, it is essential that it is free from any undue influence or abuse of power. No conflict of interest (also known as a dual commitment, competing interest, or competing loyalty) or prejudice must be allowed to influence the submission of a manuscript, its review, or the decision on whether or not it should be published (see Golden Rule 12).

What are conflicts of interest?

What is a conflict of interest, and how can it be recognized? The ICMJE, in its Uniform Requirements for Manuscripts Submitted to Biomedical Journals, gives the following guidance: 'Conflict of interest exists when an author (or the author's

---

Golden Rule 12

No conflict of interest or prejudice must be allowed to influence the submission of a manuscript, its review, or the decision on whether it should be published.

institution), reviewer, or editor has financial or personal relationships that inappropriately influence (bias) his or her actions'.[2] Basically, a conflict of interest is a conflict between private interests and official responsibilities. A simple, and helpful, guideline recommended by a number of journals is to ask the question: If certain facts were undisclosed and emerged later by some other route, would they cause you embarrassment or recrimination? It is important to realize that the potential for bias, both positive and negative, can exist whether or not someone believes that a relationship could affect their judgement. Some relationships may not have any effect; others may have the potential to influence judgement greatly. The key factor is disclosure. If people – editors and readers – know about relationships that carry the potential for conflict of interest, they can make decisions and evaluations that take these into account.

Beware!

The potential for bias, both positive and negative, can exist whether or not someone believes that a relationship could affect their judgement.

All the parties involved in the peer-review process, i.e. authors, editors, reviewers and editorial staff, need to consider if they have any relationships that may carry the potential for conflict of interest. Increasing numbers of journals are asking authors to declare explicitly if there are any conflicts of interest, and they have devised specific forms for this (see Appendix II, pages 221–231 for examples). A whole range of instructions exists about disclosure, from vague to very prescriptive. As regards reviewers, many journals ask in their guidance documents that reviewers excuse themselves from review if any conflict exists. Editors may be asked to sign disclosure statements by their journals, management bodies, societies or owners, but more often than not this is more of an understood requirement, with the expectation that editors will disqualify themselves from handling manuscripts for which a conflict of any kind exists (see Golden Rule 11). Editors should be able to decline dealing with any manuscript they feel uncomfortable handling, for example from individuals or groups towards whom they feel hostility, or where they recognize that their intellectual leanings or personal feelings may be such as to introduce a positive or negative bias.

Various types of conflict of interest can exist.

Financial
This is one of the easiest conflicts to identify and declare, but journals need to decide whether thresholds should apply in terms of both monetary amount and time since involvement before disclosure is necessary. Financial conflicts can take a number of forms.

1   *Research funding.* Receiving funds or support, for example equipment, supplies, or meeting travel and attendance costs, for a research project from a body that

might gain or lose financially through publication of a paper could be a source of conflict. Most commonly, this could involve a commercial product whose success is very dependent on whether publications on it are positive or negative. Some journals will not consider manuscripts describing research on a commercial product if the research has been supported financially by a company involved in the manufacture of that product. Those that do may require that the authors sign a declaration that they had full access to the data and its analysis, and control over the writing of the report and the decision to submit for publication (see Golden Rule 12 and this chapter, page 157).

2   *Personal financial interests.* Ownership of stocks or shares in companies that stand to gain or lose financially by publication of the work represents a potential conflict, as does the payment of consultancy fees or other forms of remuneration.

3   *Employment related.* Employment, either during the research project, at the present time or in the near future, by an organization that would stand to gain or lose by publication of the work can be viewed as a potentially conflicting situation.

Personal

Personal relationship by virtue of family, marriage or friendship is a source of potential conflict. People in family and marriage relationships may not have the same family name and so may not be easily recognized as such. However, editors' specialist knowledge of their fields of expertise will often help identify such cases and they should alert editorial staff about any that become relevant, so that certain people are not asked to take responsibility for that manuscript and inappropriate reviewers are not approached. People invited to review manuscripts in such cases should declare to the journal their relationship to the authors. Editors should not handle manuscripts from their own institutions, and definitely not from their own research groups, as they may be seen to have vested interests in seeing such papers published. Personal animosity, for whatever reason, can lead to potentially conflicting situations, and editors and reviewers should excuse themselves from being involved with manuscripts from people towards whom they feel this way. There is no need for them to give the actual reasons if they don't want to – it's enough just for them to declare this is how they feel.

Intellectual

Holding strongly opposed opinions or theories to those described in a submitted manuscript can lead to a potentially conflicting situation. If reviewers do not feel able to give a fair assessment of work because their basic intellectual beliefs will stop them entertaining contrary findings, they should declare this and excuse themselves from review. Likewise, editors may be unable to make decisions without bias. They should involve or pass responsibility to another editor if they are in any doubt, both so that the manuscript can be handled as fairly as possible and so that this is seen to be so, without an editor's own predilections affecting outcome.

Professional

All researchers in a field compete for professional advancement and possibly for jobs and promotion. If they are directly competing for research funds, or trying to beat one another to publication, they may be unable to provide an objective view, or to 'ignore' the results, new techniques, and so on, that they would have privileged access to if they reviewed a manuscript. It is a very fine balance, as genuine 'peers' need to review work but they cannot be so close that it might affect their judgement. Specialist editors will often be aware of professional rivalries, and histories, and make reviewer selections accordingly. They will be aware not only of genuine and unhealthy rivalries, but also of those that are only imagined to be so. The competitors whom authors fear most and suspect of holding back reviews or unfairly assessing their manuscripts can actually be amongst the most generous and enthusiastic of reviewers! But editors should also be wary of using reviewers who have a vested interest in someone else's paper being accepted and published in order to strengthen an area of research or to build up the activity and literature in a field to give it greater momentum (see Chapter 4, page 50).

Political, religious, racial or gender related

Any prejudices involving politics, religion, race or gender must not be allowed to affect the assessment of work and decisions that need to be made (see Golden Rule 12). The culprits may not be aware that they have these prejudices. It is up to editors and editorial staff to be alert to the possibility that they may exist and to take them into account when identifying reviewers for manuscripts.

How should conflicts of interest be handled?

Just because the potential for a conflict of interest exists, this does not mean that it will result in any wrong or biased action. It should be stressed again that the key is disclosure of any potential conflict of interest so that this information can be used in the evaluation and decision-making process. In the pursuit of transparency of the peer-review process, and in the face of increasing links between industry and researchers and the increasing numbers of start-up companies in the academic sector, journals are being urged to introduce or tighten up their policies on the disclosure of conflict of interest. Following several high-profile fraud cases, where potential financial conflicts apparently existed but were not disclosed, criticisms have been levelled against the top multidisciplinary science journals that their disclosure policies are too weak and need to be made stricter.[14,15] However, it has to be recognized that conflict-of-interest policies are hard to impose and monitor.

Concerns about financial competing interests have been most prominent in the medical field, and medical journal editors are alert to the potential influence that drug companies, for example, could exert over the publication of work they have funded. In a move designed to give power to authors collaborating with industry, the major medical journals have introduced policies requiring authors to sign statements confirming that they have been involved in designing the study and in analyzing and interpreting the data. They must also declare that they have seen the raw data and that the sponsor of the research has not had control over whether the results can be published or when they can be published (see this chapter, page 157, and Golden Rule 12). These journals will reject reports from trials sponsored by drug companies unless the authors have been granted explicit control over their data and the decision to publish. Guidelines can be found on the websites of the ICMJE (www.icmje.org), the Committee on Publication Ethics (COPE; www.publicationethics.org.uk), the World Association of Medical Editors (WAME; www.wame.org), and the Council of Science Editors (CSE; www.CouncilScienceEditors.org).

There is some debate on how authors' disclosures of financial interest should be used in the peer-review process, and whether or not reviewers should be made aware of them. What an editor does with details of disclosure of possible competing interest varies, and it is up to each journal to determine what its policy should be. Those that don't disclose them to reviewers argue that reviewers should concentrate on the science being reported, without being influenced by other factors. Those that do provide the information to reviewers argue that personal financial interests are a relevant factor that reviewers should take into account when assessing work. Whether any disclosure information is passed on to reviewers during review and the level of detail given if it is – from just that a tick box has been checked, to a full breakdown of financial interests – may depend on the subject area of the journal. In areas where commercial pressures and/or interference are a concern and where the effects on human health or public policy making on health or the environment, for example, may be affected by the results of publication, the impetus will be greater. Some journals, especially those in less vulnerable subject areas, may not require authors to submit forms on competing interest, feeling (or, rather, fearing) that it may be just another barrier to authors that will lead to them taking their papers elsewhere. All journals should, however, require that all sources of research funding and support be declared in papers; this information is most often given in the Acknowledgements.

Journals also differ in whether or not they publish disclosures of competing interests. Many argue that readers have a right to know about potential competing interests and to be able to take this into account when reading papers and interpreting results. If a journal has a policy of disclosure, readers have a right to know not only if authors have declared a competing interest and what it is, but also if authors have declined to respond to a request for this information. The journal *Nature* states at the end of its research papers whether the authors have declared no competing financial interests, whether they have declared competing interests (and then details are published online), or whether they declined to provide information on competing financial interests. Disclosure, as with many matters in peer review, relies on the basic honesty and integrity of individuals. There is a problem, however, that in some cases authors genuinely may not consider something to be a competing interest, whereas a journal would. Journals have, therefore, become much more specific in the questions they ask and the details they request. Policies of declaring potential competing interests are difficult to police, and despite rising numbers of researchers with links to industry there do not seem to be comparable levels of increased disclosure of competing interests. For example, it was reported in *Nature* in 2003 that of the 1300 or so papers published since it introduced the policy of requesting and publishing details of competing interests in October 2001, only 50 authors (i.e. just under 4%) had declared competing interests.[16] Originally the policy applied only to research papers. It has since been extended to all papers, including review articles, where authors could significantly influence people by selective or personal interpretation of a field to their benefit.

Beware!

Authors may not consider something that is relevant to be a competing interest – therefore be specific.

Moral dilemmas

There may be occasions when issues other than the subject and quality of the work come into play and impinge on the treatment of manuscripts received by journals. This should not happen and goes against Golden Rules 8 and 12.

Political or human rights issues

There have been instances where editors of journals have refused to consider papers from certain countries. For example, in 2002 it was reported that a paper submitted to a journal was returned to the author unopened, with a note saying that the journal would not accept a submission from the author's country. This 'policy' was based on alleged human rights abuses and linked to a petition calling for an academic boycott of that country.[17] After protest, the paper was considered for publication. Political issues should not impinge on academic freedom and the reporting of science. Editors who feel a conflict about this should pass the handling of such manuscripts to other editors and remove themselves completely from any involvement with those submissions (see Golden Rule 11). To refuse a submission on such grounds breaks (as well as Golden Rules 8 and 12) an important international convention on academic freedom and the principle of the universality of science: Statute 5 of the International Council for Science (ICSU, previously called the International Council of Scientific Unions), which states that the principle of the universality of science is fundamental to scientific progress.[18] The essential elements are non-discrimination and equity, and ICSU members agree to behave 'without any discrimination on the basis of such factors as citizenship, religion, creed, political stance, ethnic origin, race, colour, language, age or sex'.

Authors accused of criminal offences

Occasionally, an editor may be asked whether he or she would accept a submission from a person accused of a criminal offence unrelated to their work or serving a custodial sentence for such an offence. Editors may find that other journals have refused to consider such a submission. Any action and decision should be taken solely on scientific grounds and ethical issues related to the work (see Golden Rules 8 and 12). It is not a journal's role to make moral or character judgements on authors about things unrelated to their scientific life. Also, it can be argued that the other authors involved in the work should not be disadvantaged and their careers suffer because of the personal problems of one of the authors. In a case reported in 2001, two co-authors of a scientist accused of criminal offences not related to his work filed a complaint with a journal's publication committee when their paper was delayed as a result of debate about whether or not it could be considered or published because of the unusual circumstances. The Editor did not want to publish research bearing the criminally implicated author's name.[19] The applicants won, and after polling the opinions of the journal's associate editors a compromise was worked out whereby one of the associate editors handled the review of the manuscript. Two points of caution should be noted. Firstly, extra care needs to be taken in reviewer selection so that awareness of the crime in question does not affect a reviewer's ability to carry out an unbiased review. Secondly, be prepared for some delays in response time in such cases, as communication with authors may be subject to screening and approval by the penal authorities.

Refusals by publishers to publish articles

The decision on whether to publish an article should always reside with the editor-in-chief and his or her editors; the editor-in-chief must have full editorial independence (see Chapter 5, page 90, and Golden Rule 7). Publishers and journal owners should not exert their power to prevent publication of articles that have been accepted. Attempts by publishers to do so have led to boycotts by researchers and editors.[20]

Inability to complete review of a manuscript

It may sometimes be impossible to complete the peer-review process. For example, a proof developed with the extensive use of computers of a famous mathematical problem, Kepler's conjecture, was in review for more than 5 years after its original submission in 1998.[21,22] The conjecture, which arose because of the need to determine how best to stack cannon balls on the decks of ships in the 17th century, states that the densest possible arrangement of spheres is to stack them in a pyramid (in the same way greengrocers arrange fruit). The review process was a mammoth task, including the checking of computer code, and the many reviewers involved (a panel of 12 was used) found it impossible to certify the proof because they could not check every line of the code. Despite this, they believed, with 99% certainty, that it was correct. The editorial board of the *Annals of Mathematics* felt the reviewers had run out of energy and was therefore prepared to publish the paper, but with a cautionary note. This wasn't, however, acceptable to the author, who then went on to try to certify the proof with computers. An amended paper, containing the theoretical portion of the proof, eventually appeared in 2005 in the *Annals of Mathematics*, 7 years after the original version was first submitted.[23] This is an extreme case, and one that most editors are unlikely to encounter. However, it is an example of the right of editors to use their judgement on when to take something of a risk and to accept work for publication without certainty that it is absolutely correct because it is more valuable for the work to be out than not. This must always be done responsibly, however, and all such cases should be accompanied by appropriate notes explaining the situation and describing any limitations or uncertainties.

References

1   Wellcome Trust. Guidelines on Good Research Practice. www.wellcome.ac.uk (first published January 2002, updated November 2005). www.wellcome.ac.uk/doc_WTD002753.html.
2   International Committee of Medical Journal Editors. Uniform Requirements for Manuscripts Submitted to Biomedical Journals: Writing and Editing for Biomedical Publication. www.icmje.org (updated February 2006; accessed 25 May 2006).
3   Adam, D. and Knight, J. (2002). Journals under pressure: Publish, and be damned.... *Nature*, **419**, 772–776.
4   American Physical Society. APS Guidelines for Professional Conduct. www.aps.org/statements/02_2.cfm (updated November 2002; accessed 5 June 2006).
5   Rennie, D., Yank, V. and Emanuel, L. (1997). When authorship fails. A proposal to make contributors accountable. *JAMA*, **278**, 579–585.
6   Verhagen, J. V., Wallace, K. J., Collins, S. C. and Scott, T. R. (2003). QUAD system offers fair shares to all authors. *Nature*, **426**, 602.
7   Weller, A. C. (2001). Editorial Peer Review: Its Strengths and Weaknesses. ASIST Monograph Series, Information Today, Inc. Medford, New Jersey.
8   Cho, M. and McKee, M. (2002). Authorship in biomedical research: realities and expectations. *Science's Next Wave*, 1 March. http://nextwave.sciencemag.org/.
9   International Human Genome Sequencing Consortium. (2004). Finishing the euchromatic sequence of the human genome. *Nature*, **431**, 931–945.
10 International Chicken Genome Sequencing Consortium. (2004). Sequence and comparative analysis of the chicken genome provide unique perspectives on vertebrate evolution. *Nature*, **432**, 695–716.
11 Abbott, A. (2002). Dispute over first authorship lands researchers in dock. *Nature*, **419**, 4.
12 Over, R. and Smallman, S. (1970). Citation idiosyncrasies. *Nature*, **228**, 1357.
13 Wren, J. D., Grissom, J. E. and Conway, T. (2006). E-mail decay rates among corresponding authors in MEDLINE. *EMBO Reports*, **7**, 122–127.

14   Check, E. (2006). Journals scolded for slack disclosure rules. *Nature Physics*, 16 January. doi:10.1038/news060116-6. www.nature.com/news/2006/060116/full/060116-6.html.

15   Centre for Science in the Public Interest. (2006). CSPI calls on journals to strengthen disclosure of conflicts. www.cspinet.org/new/200601121.html (12 January 2006; accessed 30 January 2006).

16   Knight, J. (2003). Journals wrestle with definitions of 'competing' interest. *Nature*, **423**, 908.

17   Beckett, A. and MacAskill, E. (2002). British academic boycott of Israel gathers pace. *The Guardian*, 12 December. www.guardian.co.uk/uk_news/story/0,3604,858363,00.html.

18   International Council for Science (ICSU). www.icsu.org/5_abouticsu/STATUTES.htm#5 (accessed 26 May 2006).

19   Dalton, R. (2001). Journal will publish accused scientist's work. *Nature*, **409**, 548.

20   McKie, R. (2004). IBM fights to suppress cancer probe. *The Observer*, 20 June. http://observer.guardian.co.uk/international/story/0,,1234102,00.html (accessed 2 June 2006).

21   Szpiro, G. (2003). Does the proof stack up? *Nature*, **424**, 12–13.

22   News in Brief. (2004). Journal juggles balls to publish Kepler paper. *Nature*, **428**, 686.

23   Hales, T. C. (2005). A proof of the Kepler conjecture. *Annals of Mathematics*, **162**, 1065–1185.

9 Misconduct in scientific research and publishing – what it is and how to deal with it

As mentioned at the start of this book (see Chapter 1, page 4), the peer-review process depends on the trust and good behaviour of all the participants. Unfortunately, as in all areas of human activity, good behaviour sometimes falls by the wayside, and misconduct occurs. Worryingly, the incidence of misconduct in science appears to be increasing. Or perhaps it is just being picked up more frequently or more people are being made aware of it. Whichever is the case, most editors and journals can, unfortunately, expect to come across instances of suspected or alleged misconduct. The misconduct may range from the relatively minor to the very serious, and it may involve any of the parties in the peer-review process. Whatever the suspected or alleged type and extent of misconduct, there is one absolute and overriding rule: suspected or alleged misconduct must not be ignored (see Golden Rule 13). It is the duty of editors-in-chief and journals to look into each case at journal level and decide whether there is any substance to the suspicion or claim of misconduct, and then either to deal with it themselves or to alert the appropriate agency for further investigation and action. This applies both to submitted manuscripts and to papers already published. Editors must resist any temptation just to reject a manuscript that they suspect may be problematical and to pass it, and so the problem, on to another journal. This not only opens up the risk of inaccurate or fraudulent data being published and becoming part of the literature, it may also lead people who misbehave to think that, if they can get away with it once, why not again?

Editors and journals must, however, always be very careful to distinguish between genuine errors and the intention to deceive; the latter constitutes misconduct, the former does not. They should not make or spread allegations of misconduct before investigating any suspicions and finding them to be substantiated. Not only can this seriously, and perhaps unjustly, damage a researcher's reputation, it also opens up the possibility of litigation. It must be recognized that some apparent 'misbehaviour' may be the result of ignorance of good practice. This can happen, for example, with authors who do not have much experience of research or publishing, or with very junior authors who have not received much, or perhaps any, guidance from their supervisors in these areas. It is expected that, as part of the scientific process, junior

---

Golden Rule 13

Suspected or alleged misconduct must not be ignored.

researchers gain an understanding of the practicalities and ethics of research work and scientific publishing from their supervisors and this should form part of their research education. However, there is evidence to suggest that this is not always so. In a survey of its junior researchers, the American Physical Society found that a significant number of physics students complete their doctorates without any formal instruction in data collection and recording.[1] Many of the respondents called for more attention to be given to ethics questions. They felt that, although proper mentorship by supervisors was key to establishing patterns of ethical behaviour, there should also be mandatory ethics seminars and discussion. Journals also have a role to play in this educative process, by publishing guidelines, providing feedback and highlighting inappropriate or bad behaviour. If misbehaviour due to inexperience or inadequate knowledge on the part of junior authors is found in manuscripts submitted to, or in papers published by, a journal, the senior authors or principal investigators must also accept responsibility for this, as they have failed in their supervisory and instructor roles; they cannot plead exemption due to ignorance of what people in their research groups are doing or submitting for publication. They also, therefore, have to accept the penalties that journals may impose.

Beware!

Apparent misconduct may be the result of genuine error. Take great care not to make or spread allegations that have not been substantiated.

Editors are very busy people. For many, their editorial duties represent just a small part of their professional activities, and some editors may be working with very little support and possibly inadequate resources. Investigation of alleged misconduct can be very time-consuming. The situations are frequently sensitive, calling for diplomatic handling, and there may also be legal implications. It is therefore not surprising that some editors may be reluctant to initiate investigations, even at journal level, despite being ethically obliged to look into all problematical issues. There may, however, be editors who are not aware of the sorts of problems that can arise, what can be done to address them, and where they can turn for help. This chapter provides some guidance on this.

What types of misconduct can occur?

Author misconduct

In 2005, the journal *Nature* reported a survey in which scientists funded by the National Institutes of Health (NIH) in the USA were asked about misbehaviours relating to their work.[2] Over 3000 scientists who were in either early or mid career were surveyed in 2002 and asked about their work-related behaviour over the previous

3 years. The misbehaviour categories presented ranged from very serious to relatively minor. Included amongst the more serious were: falsifying data, using other people's ideas without permission or due credit, failing to present data that contradicted their own research, overlooking others' use of flawed data or questionable interpretation of data, and changing the design, methodology or results of a study in response to pressure from a funding source. Not only was a wide range of questionable behaviour found, the numbers of scientists admitting to such behaviour were worryingly high. The situation becomes even more worrying when one considers that the figures obtained were probably conservative as there may have been under-reporting due to concern by the respondents about possible identification even though their anonymity had been assured. Despite this possible concern, around a third of the scientists admitted to some form of misbehaviour in the top-10 most serious categories during the previous 3 years. The authors of the report felt that the occurrence of the types of misbehaviour they found represents a greater threat to the integrity of science than that resulting from the high-profile cases of fraud that appear in the media (see this chapter, page 176). (As an aside, it should, however, be recognized that the cases of scientific fraud that do hit the media are very damaging as they seriously undermine public confidence in science and its reporting.) For the categories of misbehaviour most relevant to scientific publishing, the results showed that 6% of the respondents had failed to present data that contradicted their own previous research, 5% had published the same data or results in two or more publications, 10% had inappropriately assigned authorship credit, 11% had withheld details of methodology or results, and 15% had dropped observations or data points from analyses. Very alarmingly, nearly 30% admitted to inadequate record keeping related to research projects.

This survey shows that not only is there a real problem with research misbehaviour, but that it is quite a significant one. Editors therefore need to be alert to this. It would, however, be totally unrealistic to expect editors and the peer-review process to be able to detect many of the potential forms of misbehaviour. For example, it would be impossible to know if data points had not been included or experiments had not been carried out in the way reported. Editors and reviewers have to assess what is submitted. They can, however, and should, always request more data or clarification if they come across anything that is unclear or suspicious. Part of their role in the peer-review process is to pick up errors and ensure that these are corrected or addressed before work is published.

Why is there this relatively high level of misbehaviour? One can speculate that it is partly linked to the great pressure on researchers in an increasingly competitive market to publish in order to get funding and promotion. There is also pressure to be the first to publish, as that brings the greatest rewards, and so corners may be cut and standards lowered to achieve this. Part of the blame has also been levelled at the top multidisciplinary science journals, which have been accused of rushing manuscripts through review in their quest to publish the hottest papers and to be the first to do so.[3] There are concerns that standards may be being compromised. These journals, which do receive exceptional and highly novel and innovative manuscripts, must also ensure that the review of such manuscripts is exceptionally rigorous and extra-stringent. Reviewer choice is absolutely critical, and the number and range of reviewers used is likely to be greater than usual, particularly for manuscripts reporting unexpected results or if previous findings or hypotheses are being overturned.

What is scientific misconduct by researchers? The Office of Research Integrity (ORI; http://ori.dhhs.gov, and see this chapter, page 189) in the USA defines it as follows:

> 'Scientific misconduct' or misconduct in science means fabrication, falsification, plagiarism, or other practices that seriously deviate from those that are commonly accepted within the scientific community for proposing, conducting, or reporting research. It does not include honest error or honest differences in interpretation or judgments of data.

And it defines a suspect manuscript as one

> submitted to or published in a journal which is suspected of including or being based upon falsified or fabricated data, results, or methodology or plagiarised text or ideas.

What is meant by these categories of misconduct and what types of behaviour fall into those categories?

Fabrication and falsification
Fabrication is where data or facts are just made up or invented, and falsification is where data or facts are altered or manipulated. They can range from the creation or alteration of a single datum point to the invention of a whole series of experiments or manipulation of data on a massive scale. Deception, dishonesty and false representation are involved, and therefore these activities constitute fraud. The behaviour is not accidental; there is always the intention to deceive. There have, over the years, been significant numbers of cases of scientific fraud reported.[4] Whereas the majority may not have been very widely known outside of the scientific community, the whole world has been rocked by a number of very serious fraud cases in the first years of the 21st century. These have been reported extensively in the popular media, and so have become highly visible. That the work involved was published in some of the top and most prestigious science journals after peer review has sent out worrying signals not just to the scientific community but also to the general public. The more such cases there are, the more public confidence in the scientific process is eroded. The following two cases highlight such scientific fraud.

The Jan Hendrik Schön case. In 2002, Jan Hendrik Schön was acknowledged to be a brilliant young physicist. He had a prolific publication record, with many of his papers published in the top science journals. However, a number of groups were having problems repeating his experiments and questions began to be asked about the validity of his work. It was also then noticed that he had presented identical graphs in different publications to show the outcome of supposedly different experiments. An investigation committee reporting in September 2002 found him guilty of falsifying or fabricating data in 16 of the 24 alleged cases investigated.[5] Fifteen of these papers were published, and subsequently retracted, in the journals *Science* and *Nature* (see this chapter, page 195). This was fraud on a truly massive scale. Not only was Schön totally discredited, the review processes of the journals involved attracted much criticism, especially as the reviewers hadn't spotted something as simple as duplicated figures. Fuller details of the Schön case can be found in references 3 and 6.

The Woo Suk Hwang case. In 2004 and 2005, the stem-cell researcher Woo Suk Hwang published two landmark papers in which he claimed, respectively, to have cloned human embryos and harvested stem cells from them,[7] and to have established 11 patient-specific embryonic stem-cell lines.[8] These were hailed as remarkable achievements, of great importance for therapeutic cloning. The significance of the work resulted not only in great professional recognition for Woo Suk Hwang, but also in considerable personal glory and public acclaim. Within a very short time, however, doubts were raised both about certain ethical issues concerning egg collection to provide material for the experiments and about the work itself. The work in both papers was very soon discredited and Woo Suk Hwang was found guilty of massive fraud and ethical transgression and the papers were retracted (see this chapter, page 195). A timeline of events shows how rapidly all this occurred.[9]

Schön and Hwang were both found to be guilty of fraud on a massive scale. However, in both cases the fraud came to light relatively soon after the publication of the fraudulent papers. They are therefore also good examples of the self-correcting nature of the scientific process.

Plagiarism
Plagiarism is the taking and use of others' ideas, writings and inventions, and passing them off as one's own, i.e. without giving credit to the originator. It is a misbehaviour that is becoming more common in many areas of life, from school through college to research. One of the main reasons for this is that there is now an enormous amount of information available via the Internet; text is very easy to copy and paste, and ideas can be gleaned from a multitude of sources. Ideas and content can also be taken from privileged information, such as that available to editors and reviewers in both scholarly publishing and grant reviewing. The extent of plagiarism can be very wide, from copying a few phrases to wholesale duplication of text and experiments.

Anti-plagiarism software is available, and many colleges and universities routinely scan students' work to check for similarity to existing documents and to the work of others in the peer group. The publishing industry is looking at ways to adapt the technology to the checking of manuscript submissions.[10] The ideal solution would be a tool that was part of online submission and review systems and which would automatically check each submission, alerting the editor or editorial office to suspect papers for further investigation. However, one great problem with anti-plagiarism software is that it can only check against what is freely available. It cannot therefore check against publications that are subject to subscription controls or, importantly, against work that may have been submitted and be under review elsewhere. Co-operation between publishers to allow cross-checking of their archives would be one way to help resolve the first problem; the second problem is much harder to deal with, as manuscript submissions must be kept confidential by the journals to which they have been submitted (see Golden Rule 3). Until some industry-wide tool becomes available, journals need to consider if they have the time and resources to submit their submissions, or a random selection, to some plagiarism-detection step.

Decisions also need to be made on what is to be classified as plagiarism in each individual case and this requires some common sense. Duplicating text or results from other articles or books is clearly unacceptable, either from the work of other authors or from an author's own (known as auto- or self-plagiarism). But there are only a certain number of ways of describing some methods and techniques. Also, certain strings of words may be very common in some disciplines. This isn't plagiarism. On the other hand, it is almost impossible to spot some serious cases of plagiarism – for example, those cases where the ideas and arguments of others have been used but have been rephrased from the original, or expressed in a very different way, so as to make links undetectable by anti-plagiarism software. Useful information on anti-plagiarism software and good examples of what is and is not considered plagiarism are available.[11]

Redundant or duplicate publication

Redundant or duplicate publication occurs when a paper is published that is either the same as one already published or that contains significant overlap with it. This may occur more than once, so that a number of identical or similar papers exist in the literature. Not only is this unethical, it also distorts the significance of the findings and leads to inaccuracy in meta-analyses. What is the extent of this problem? Studies carried out in the areas of anaesthesia/analgesia and surgery indicate that around 5–11% of articles may be covertly identical or almost identical (i.e. without referencing the original article), and up to nearly a quarter may have some form of redundancy.[12,13]

Submitting authors need to be aware that different journals and publishers have different rules about what is acceptable regarding prior publication of results. Different journals with the same publisher may also have different rules (for example, *The Lancet* and *Cell* compared with other journals at Elsevier at the time of writing in 2006). This is a very complicated area. For some journals, any prior communication, such as publication in conference proceedings or posting on a public server, disqualifies a paper from consideration. As stressed in Chapter 3 (page 26), journals must make very clear what their policies are on this. It is also crucial that all editorial staff understand what the policies of their journals are so they can pass on the correct information to potential authors if they ask.

Dual or multiple submission

If authors submit the same manuscript to more than one journal at a time, this is called dual or multiple submission; this is not ethical and should not be done. No author should submit the same work simultaneously to different journals to ensure publication somewhere by a certain time, for example if trying to beat competitors to publication, withdrawing from the other(s) once a journal has accepted the manuscript for publication. This is one of the reasons why journals should request an explanation from any author who withdraws a manuscript after submission. If there is any suspicion that an author may have been guilty of dual submission prior to withdrawing his or her manuscript, the literature should ideally be monitored subsequently to see whether the paper appears in another journal. If it does, the dates of submission and acceptance should be checked to see if they overlap with the time the manuscript was under consideration by the journal from which the manuscript was withdrawn. I used the words 'should ideally be monitored' above, because in reality many editors and journals will not have the time or resources to carry out such searches, even if the number of suspected instances of dual submission is low.

Digital image manipulation

The arrival of digital photography and image-manipulation software means that images can now very easily be altered or enhanced. The basic rule is that any digital technique has to be applied to the whole image; selective contrast enhancement of one area, for example, is not allowed. Authors must not use image-manipulation software to inappropriately enhance or misrepresent their results. However, although there will be some authors who intentionally misrepresent their results and/or their significance in this way, there will be others who may make inappropriate enhancements without realizing that what they are doing is wrong. In an attempt to present their results optimally, researchers may unknowingly cross the line between what is acceptable enhancement and what is effectively misconduct.[14] To prevent any misunderstanding, authors need guidance, and journals are beginning to provide advice to their authors. The *Journal of Cell Biology* (www.jcb.org) is at the forefront of this effort and has introduced stringent policies on what is not acceptable. Readers whose journals feature images that are candidates for enhancement are advised to consult the guidance given by that journal.[15] With the advent of electronic submission and publication, many journals require authors to submit their figures as electronic files as part of a totally electronic workflow. This presents the opportunity for these files to be checked for any evidence of manipulation, something that would not have been possible when only paper copies were being submitted. The *Journal of Cell Biology* takes this issue very seriously and has specially trained editors to check images and to look for inappropriate modification in all of its accepted papers before they are published. During the first 3^{1}/$_{2}$ years of screening, the *Journal of Cell Biology* found that only 1% of its accepted papers contained fraudulent image manipulation, but that 25% had at least one figure that had to be redone because of the presence of manipulation that violated its guidelines, suggesting that there may be considerable lack of understanding about what is and what is not appropriate manipulation.[16] The ORI also has guidance on how to detect image fraud using software 'Droplets' and gives step-by-step instructions on how to do this.[17] Journals for which image manipulation might be a real problem may want to request that authors submit original images with their 'optimized' ones and/or list the changes they have made to all the images, and that they name the image-processing program used. In cases of suspected image-manipulation misconduct, the authors should definitely be asked to provide the original, unenhanced images if they haven't already done so, and additional images in support of the work if necessary.

Researchers may also unknowingly be using images that were subject to error at the time of preparation because of the condition of the samples or how the images were captured. Imaging technology has become very sophisticated, but there is frequently insufficient understanding of how it should be used and how errors may be introduced. Alison North has put together an excellent guide for beginners on the practical pitfalls in image acquisition and how they can affect data interpretation.[18] This article also contains a highly useful and informative online bibliography of resources available to readers wanting to pursue this in greater depth. Editors whose journals publish work where biological images are an integral component of the scientific evidence will do their communities a great service by pointing authors to guidelines such as these.

Authorship problems
Various authorship problems can arise and these are described in Chapter 8 (see pages 148–158) along with guidance on the establishment of authorship and the ethical obligations of authors (see also the Good Practice Checklist in Appendix I). Deviations from ethical conduct can be considered to be misconduct, for example gift and ghost authorship, intentional failure to include all authors, the inclusion of fictitious names, or the forging of the signatures of other authors.

Failure to disclose conflicting or competing interests
The issue of conflicting or competing interests is covered in Chapter 8 (see page 164). If a journal has a requirement for authors to disclose conflicting interests when they submit to that journal and an author fails to do this, then that is misconduct. If the journal has listed the things it considers to represent a potential conflict then authors cannot plead ignorance of them. However, if a journal's policy is not made clear, authors may declare that they were unaware that their particular circumstances represented a potential conflict that should have been declared. Appendix II gives examples of some conflict-of-interest forms in use, but journals should design forms that are appropriate to their disciplines and communities, and in line with the potential level of gain that might be achieved by non-disclosure. For example,

journals in fields where the financial rewards are significant if products or techniques are endorsed by publication need to be extra careful.

> ## Beware!
>
> Beware inadequately describing what are considered to be potential conflicting or competing interests that should be declared by authors on submission.

Misrepresentation of personal communications

Some authors may intentionally either misrepresent personal communications they have received from other researchers for their own benefit or include more than those researchers intended be made public, including personal correspondence. This is not ethical and communication with the person being cited may be necessary or insistence that signed approval must be submitted by the authors if it hasn't been provided as part of the submission requirements (see Chapter 3, page 30).

Failure to abide by the policy requirements of a journal

Different journals have different policy requirements that are relevant to their subject areas or editorial policies and that they expect their authors to follow: for example, acknowledgement of funding sources (although this should be universal), the provision of written approval for personal communications, the submission of ethics committee approvals, the provision of accession numbers, the deposition of data or experimental material in public repositories, and the free distribution of materials used in papers to other researchers for non-commercial purposes. Some non-compliance issues are apparent on submission and can be sorted out then or during review, and acceptance for publication made conditional on their resolution. Others, however, such as failure to supply materials, will not come to light until after publication. Common sense and understanding need to be brought into play when dealing with these to establish whether there has been any transgression or if there is some simple explanation; for example, researchers not getting responses from an author because the author has moved and they've been using old contact details for that person (see Chapter 8, page 158). For policies of free distribution of materials there are also resource considerations, as supplying materials can be both time-consuming and expensive, especially if a large number of requests are received by a research group following publication of a paper. The ideal solution is for materials to be deposited in public repositories or stock distribution centres, which will then take over the responsibility for fulfilling requests for materials. Unfortunately, there aren't repositories in all areas, and there are funding problems in some existing ones. Materials provision can also be complicated by legal issues connected with commercial company involvement and misunderstanding by the authors on what they are allowed to do with certain materials. Readers are referred to the article by David Cyranoski for a good summary of the issues involved in material availability.[19]

Reviewer misconduct

Reviewers are the people against whom most accusations of misbehaviour are levelled. They are also often the people who are criticized when peer review is denigrated; detrimental comments are made about their lack of integrity, slowness, lack of expertise, and so on. Authors may, if the review process appears to be proceeding too slowly, start to suspect that one of the reviewers is intentionally holding up the process, or, if a negative or critical review is received, that it is from a competitor acting unfairly. Yet these individuals are experts in their fields and regularly give freely of their time and expertise, a situation rarely encountered in professional life. Reviewers are the cornerstone of peer review and without them it could not survive. They are a valuable resource and crucial to the standing and success of any journal (see Chapter 7). But they are human, and as for the other groups involved in the peer-review process, can lapse into bad behaviour. What types of misconduct can reviewers be guilty of?

Failure to disclose conflicting or competing interests
All reviewers should declare if they have a potential conflicting or competing interest when they are asked to review a manuscript (see Golden Rule 11) and it is up to the editor to decide whether or not to use them (see Chapter 4, page 58). They should not review a manuscript they know they cannot assess fairly, doing so only to be able to submit a damning review or just to get sight of the information within that manuscript.

Disclosing confidential information to others
One of the Golden Rules of peer review is that everyone involved in the peer-review process must keep all information associated with the submission and review of a manuscript confidential (see Golden Rule 3). This applies to reviewers and they should resist all temptation to talk about the manuscript with others or leave copies of it lying around where others will see it.

Using authors' ideas or results
Reviewers must not take any ideas or results from manuscripts they review and attempt to pass them off as their own or use them for their own benefit in any way (see Golden Rule 10).

Delaying or not returning reviews
Some reviewers will take longer to review than they originally anticipated, either because of shortage of time or the need to deal with other commitments, or because a manuscript turns out to be more complicated than they'd envisaged. This is totally fine, although it is always helpful if they can let the journal know that their review will be delayed. A reviewer should not, however, intentionally delay submitting a review, either because they don't take the commitment seriously or because they have an ulterior motive – for example, because they want to see their own work, or

that from another group, published first, i.e. they are effectively 'scooping' the authors of the manuscript they are reviewing. Failure to submit a review may also be due to this, or it may be due, again, to a lack of commitment and responsibility. Both are wrong, but the former is much more serious.

Damaging the character or reputation of another scientist
Reviewers must not make damaging personal or derogatory comments about the authors whose work they are reviewing – they should concentrate on providing an objective assessment of the work (see Golden Rule 6). They should not make such remarks in either their comments for the authors or in their confidential comments to the editor. In the latter case, the author has no opportunity to rebut any negative comments.

Editor misconduct

So far, all the misconduct described has been on the part of the authors or reviewers. However, editors can also be guilty of misconduct. The Committee on Publication Ethics (COPE; see this chapter, page 188) has introduced a code of conduct for editors to give them guidance on being fair to authors, researchers and readers.[20] This code is aimed at biomedical editors, but much of it applies to editors from all disciplines. Deviations from the code are considered to be misconduct. Chapter 8 in this book (pages 158–162) describes the obligations and responsibilities of editors, and negligence in a number of these can be considered to be misconduct.

The main types of editor misconduct are:

- abusing the trust of the parties involved in the peer-review process
- acting in a defamatory way when carrying out editorial duties
- not respecting the confidentiality of information obtained during submission and review
- using privileged information for personal gain or to disadvantage or discredit others
- not declaring conflicting or competing interests, or allowing these to influence actions and decisions, either positively or negatively
- intentionally delaying the progress of a manuscript through review
- allowing decisions to be influenced by outside people or organizations, or to be made in the interests of personal gain or for commercial reasons
- accepting for publication work known to be fraudulent, erroneous or unethically obtained
- publishing their own work without proper peer review
- allowing actions or decisions to be influenced by bribes or inducements
- using funds or resources inappropriately
- using unethical means to increase their journal's Impact Factor.

Editors are not superhuman – they are mere mortals. They are therefore prone to all the usual human failings, such as forgetfulness, slowness, confusion, or lack of organization. Unless these factors are introduced intentionally, with the aim of personal gain or for ulterior motives, they should not be classified as 'misconduct'.

Genuine editorial misconduct may be difficult to pick up, but there have been some extreme examples. There is, for instance, the classic case of Cyril Burt, who founded and was editor of the *British Journal of Statistical Psychology*. Burt was found guilty of several cases of misconduct: he published a great many of his own papers; he changed the work of others without their permission; he added favourable references to his own work; he published a letter and a response, both his own, under pseudonyms so that he could criticize a colleague. This is a case of an editor massively abusing his position of trust and getting away with misconduct for some time. The COPE 2003 report[21] gives examples of various types of unacceptable behaviour by editors, with details of how they were dealt with by COPE. The organization considers complaints against editors that have been through a journal's own complaints procedure and if the journal is a member of COPE. It deals only with the process of editorial decisions, not the substance; that is the responsibility of each individual journal.

How should cases of alleged or suspected misconduct be handled?

Journals and editors will become aware of possible misconduct in a number of ways. Some they may discover themselves, for example during the review process as facts come to light – either from a single source or as a result of putting together various bits of information from different sources. They may be alerted to something suspicious or to genuine misconduct by a reviewer, author or reader. It is hoped that not too many editors will be at the receiving end of anonymous letters written in capitals as reported by the *Journal of Clinical Investigation*![22] If any suspicions start to form, it is vital that complete and accurate records are kept of everything. Comprehensive records should, of course, be kept for the peer review of all manuscripts, but it is especially important in cases of suspected misconduct. They also need to be kept confidential and information restricted to the minimum number of people – just those associated with the investigation – to avoid damaging an individual's reputation and incurring possible legal action. A good rule to follow is to seek advice if in doubt or before doing anything that might result in litigation.

Although every allegation or suspicion of misconduct needs to be looked into at journal level, it is not the role or responsibility of editors to mount full-scale investigations into allegations of research misconduct that involve agencies outside the journal's activity – that is the role of the institutions employing the authors and their funding agencies. But all journals should have some mechanisms or policies in place on how to deal with such problematical issues at journal level. Each allegation or suspicion needs to be dealt with on a case-by-case basis and action decided accordingly. There will always be unique and possibly extenuating circumstances, and so it is difficult to stipulate exactly what should be done. However, the following schedule is offered as general guidance in cases of alleged or suspected misconduct.

1 Treat every allegation of suspected misconduct seriously and look into it to see if there is any evidence to support it. All journals should have procedures in place for investigating allegations of misconduct. Initial preliminary investigation may be by the editor-in-chief or managing editor (unless the accusations are against these people), with referral to a select group if the initial investigation shows that there may be a case to answer. The composition of the group will depend on the size and structure of the journal. If the journal is one of a group owned or run by a society, it could, for example, comprise all or some of the following: the editor-in-chief, the managing editor, the editor(s) associated with the case under investigation, the publications director or officer, a member of the publications committee, a society officer, and perhaps an editor-in-chief or appropriate editor from one or more of the other journals. If it is an individual journal, the group may comprise the editor-in chief, the managing editor and the most appropriate members of the editorial board. But group composition will, of course, depend on the editorial structure of the journal(s). The journal's publisher, if there is one, should also be able to offer advice and assistance, particularly if it is a large publisher, as one of its other journals may well have had to deal with a similar case. The publisher may already have guidelines in place. Help is also available for those editors who feel out of their depth or have no one to turn to for advice (see page 187 below). If the accusations of misconduct are against the editor-in-chief, the complaint will need to be referred to the society, other owner, or publisher, as appropriate.

2 Keep a written record of all evidence and all communications, with dates. Dates are very important and may be needed, for example to determine priority. Go over the whole history of events very carefully when the investigation starts and compile an accurate summary document containing dates, significant facts, and an audit trail of the people involved at the various stages of the review process and in the decision making. This chronological log will be invaluable when re-visiting the file, which may be necessary a significant number of times, and over a considerable period, in some investigations.

3 Keep everything confidential – do not involve more people than are necessary. Premature release of information, or release to inappropriate people, may damage the reputation of a person who may subsequently be found to be totally innocent of misconduct. A person who finds themselves in this situation may decide to sue a journal or publisher for damage to their professional reputation and future prospects. Do not put the responses from the various parties involved in the investigation onto the manuscript details as they come in if an online submission and review system is used by the journal. Do this when and if appropriate – normally at the end of the investigation – and take care as to what information and attributed comments you do put online.

4   It is of utmost importance in all cases of suspected misconduct to ask the person against whom an allegation has been made for a response and an explanation. They must be given this opportunity. It may in many circumstances be wise to keep the source of the allegation confidential and use terms such as 'It has been brought to our attention that . . .'. Do this in writing, present the case clearly and ask for specific responses and an explanation, in writing and by a specified date. Written evidence should also be obtained from the other parties involved. One person should take responsibility for dealing with the correspondence, writing on behalf of the group if more than one person is involved in investigating the allegation.

5   Where it looks likely that allegations will prove to be founded, for example where it is clear after investigation that there has been plagiarism or dual submission of a manuscript, the review process should be stopped and put on hold if the situation relates to a manuscript under consideration for publication, and the appropriate parties (editors, reviewers, editorial office staff, depending on the circumstances) notified of the reasons for this. The authors should still be required to provide a response. If the allegations are against the editor or reviewers, those individuals should be removed from involvement with that manuscript and an alternative editor or new reviewers should be found. If definite evidence of dual submission is found – for example if a reviewer contacts the journal to say that he or she has received the same manuscript for review from another journal (see Chapter 4, page 59) – contact with the editor(s)-in-chief of the other journal(s) will be needed, even though this breaks Golden Rule 3, which states that the submission of a manuscript and all the details associated with it must be kept confidential. This needs to be done carefully, after it has been verified that the manuscripts are the same (i.e. in content and it is not just that the titles are the same), and in confidence. Both (or all) editors-in-chief should then contact the author to request an explanation as to why their journals have the same manuscript submitted for consideration for publication. It is not always the case that there has been misconduct, and dual submission may have been the result of confusion (for example, as described in point 6 below). If plagiarism or dual submission are discovered only after a paper has been published, measures need to be taken to correct the literature (see this chapter, page 192) in addition to any punitive measures that might be taken (see this chapter, page 190).

6   In the case of alleged author misconduct, if no response is forthcoming, the authors should be notified that if a response is not received by a certain date the review process will be terminated, the manuscript will be returned and further action, such as a ban on future submissions from the authors or notification of their home institutions, may be taken. That, particularly if the correspondence is copied to all the authors rather than to just the corresponding author, will generally bring a rapid response. Great care must be taken, however, when copying in others that allegations about specific people are not made, as if they prove

to be unsubstantiated this may result in those people taking legal action for defamation of character, damage to personal and professional reputation, and so on. The correspondence should not be accusatory, but rather should state the facts, outline the problem, and list the responses required. Be cautious, though, that the alleged misconduct hasn't arisen through misunderstanding. For instance, dual submission may have resulted from confusion on the part of the authors about the actual status of their submission. They may, for example, mistakenly believe their manuscript has been rejected by a journal because they have looked at the outcome online and mixed up their current submission with an earlier submission with the same title, which appears as 'rejected' on the online site. They therefore go on to submit the manuscript that is still under review at the first journal to another journal, resulting technically in a dual submission. In another scenario, authors may, wrongly, have prematurely and informally been given notice of a rejection decision (see Chapter 4, page 85). After editorial discussion, this is changed to acceptance but the authors have in the meantime quite innocently gone on to submit the manuscript to another journal, again resulting technically in a dual submission. In many such cases, things are not clear cut and you will need to decide whether or not to give authors the benefit of the doubt.

7   The responses and explanations received from the individual(s) against whom the allegations have been made need to be evaluated by the investigating person or group. If it is clear that the allegations are unfounded, the review and editorial processes can be continued if these were halted during the investigation. An appropriate apology needs to be made to the individuals involved, but with the explanation that all allegations or suspicions have to be investigated by the journal. Reassurance can also be given that confidentiality has been maintained and the information kept restricted to the minimum number of people. If the allegations are substantiated, the journal needs to decide what punitive measures it will take and whether referral to the individuals' institutions and/or funding bodies is warranted (see this chapter, page 190).

The above is a scheme that outlines the general principles of how to proceed in cases of alleged or suspected misconduct. Readers who are interested in seeing 'real' schemes for handling alleged misconduct by the various parties involved in peer review can find examples in the procedures of the American Society of Plant Biologists.[23]

Where can you turn for help?

There are various organizations to which journals and editors can turn for advice in cases of suspected misconduct. They may need to do this because of their own inexperience in dealing with misconduct, or when they come across new, difficult or sensitive situations. What are these organizations?

The publisher

If the journal is published by an external publisher, they should be contacted for advice. They will very likely have experience of misconduct issues and will be able to provide or find advice, and alert editors to potential pitfalls and risks. They will also be able to advise on legal issues and whether legal advice needs to be sought.

Professional bodies

The journal's community may be represented by a professional body and this can usually be contacted for advice and guidance on all sorts of matters, including misconduct. Many professional bodies have a code of conduct that they expect their members to follow and they will be able to advise when that code has been breached. The professional body may itself want to investigate and impose appropriate sanctions if misconduct is proven.

Scholarly publishing organizations

There are a number of scholarly publishing organizations to which editors and journals can look for advice. Appendix III lists the principal ones. Many have a lot of valuable information and resource material mounted on their websites, with much being accessible to non-members. Many of these organizations also run online discussion listservs where members can post questions or ask for advice. If you do take part in any such discussion groups or post a request for help in a possible misconduct case, be sure, for legal reasons, always to anonymize cases and not include any details that might lead to identification of the people or organizations being discussed.

Beware!

When posting messages connected to suspected or alleged misconduct on listservs, do not include any details that might lead to the identification of any of the individuals or organizations being discussed.

The Committee on Publication Ethics (COPE)

The Committee on Publication Ethics (COPE) is a voluntary organization founded in the UK in 1997, originally as a self-help group set up by a group of biomedical editors for editors, to address breaches of research and publication ethics. Its remit is also to develop good practice and it has produced 'Guidelines on Good Publication Practice' which are intended to be advisory rather than prescriptive. They are available on its website (www.publicationethics.org.uk) and COPE encourages their dissemination. Editors and journals should take a look at these for general guidance and practical advice. COPE also aims to find practical ways of dealing with problems and will advise on cases of misconduct but can, for legal reasons, do so only for anonymized cases. The case studies that have been considered by COPE are given on its website; each case is presented, the discussion and advice summarized, and the outcome given if available. This is a valuable resource, not only for people seeking advice on what to do in specific cases, but also because it gives an idea of the sorts of problems and types of misconduct that can occur.

For many years, COPE campaigned for the establishment of an agency in the UK to be responsible for research integrity and to which allegations of research misconduct could be referred. That dream was realized in 2006, with the formation of the UK Panel for Research Integrity in Health and Biomedical Sciences.[24] Although the remit of this agency on establishment was to cover research in the health and biomedical sciences only, its aim is ultimately to cover misconduct in all scientific fields.

The Office of Research Integrity (ORI)

The Office of Research Integrity (ORI; http://ori.dhhs.gov) is part of the Office of Public Health and Science within the Department of Health and Human Services in the USA. It was established in 1992 as an independent office within that service (from its set up in 1989 as the Office of Scientific Integrity) and oversees the institutional handling of misconduct allegations in research funded by the US Public Health Service. However, it produces a guidance document and advice that will be useful to editors and editorial staff when dealing with suspect manuscripts.[25] The document suggests helpful editorial policies and covers the roles of editors in responding to scientific misconduct. It also lists procedures for handling suspect manuscripts and for correcting the literature if necessary. ORI will advise on research funded by the Public Health Service at all stages from the beginning of review, through review and post publication; for research funded by other bodies it will point people in the right direction and provide appropriate contact details. ORI maintains confidentiality during all its investigations and no information is released, even on which cases are being investigated, until those investigations are closed. If misconduct is found, ORI makes a public announcement.

International Committees for Scientific Misconduct

A number of countries around the world, for example Denmark, Finland and Germany, have bodies that deal with scientific misconduct and/or publish guidelines dealing with scientific misbehaviour. They vary in size and remit. A list can be found on the ORI website and readers are advised to consult that for up-to-date information (http://ori.dhhs.gov/international/websites/index.shtml).

Funding agencies

Many funding agencies now require institutions that receive funding from them to establish written policies and procedures for handling misconduct allegations. Some may insist that there must be someone in place whose responsibility it is to look into any problems. They also frequently stipulate the timeframe for the investigation of allegations and for the submission of the outcome of those enquiries. The funding agencies themselves have produced guidance documents outlining the standards they expect their researchers to adhere to and to help institutions determine if these have been breached: good examples are those of the National Science Foundation (NSF) in the USA,[26] the Medical Research Council (MRC) in the UK,[27] and the Wellcome Trust in the UK.[28] It is recommended that editors visit the website of the funding agency for the research in the suspect manuscript for advice on procedures for lodging suspicions of research misconduct.

What sanctions can be imposed as a penalty for misconduct?

If allegations of misconduct prove to be founded, the guilty party or parties should be reprimanded or penalized in some way. The sanctions imposed and their severity will depend on who is guilty of the misconduct and on the degree of misconduct.

Authors found guilty of misconduct

If it is clear after investigation that an author or group of authors did not intend to mislead or deceive but that the situation arose through either ignorance of publishing ethics or confusion (see this chapter, page 187), then a formal letter from the editor or journal to all the authors giving details of the breach of ethics that has occurred and stating what the accepted standards are, together with information on where guidelines can be found, is the most constructive solution. This provides an educative service and also acts as a wake-up call to the senior authors or principal investigators, who may be failing in their supervisory roles in the research and publishing education of their group members. Journals should keep a note of authors who have been involved in such cases. If they act in the same way again, ignorance cannot be used as the defence.

If, after investigation, an author is found guilty of misconduct, for less serious cases (for example, copying a small bit of text from someone else's paper, failing to abide by one of the journal's policy requirements), a letter of reprimand from the journal may be sufficient, with a warning that a harsher penalty will be imposed if the author is found guilty of any misbehaviour in the future. In serious cases (such as fabrication or falsification of data, appropriating someone else's results or ideas for personal, professional or financial gain), the journal or editor may decide that referral to the authors' home institution(s) or funding body is necessary, with a request that the matter be investigated and that the journal be informed of the outcome. Referral to an author's institution or funding body is a serious step, with potentially serious implications for the individual, so editors have a duty to ensure that this action is warranted in those cases that are referred. Notification should be done in writing and just the facts related. No journal correspondence should be sent. The institution or funding body concerned will need to contact the relevant parties and request whatever information it requires. Some investigations may lead to criminal proceedings so it is essential that everything is done very carefully and appropriately, and that legal advice is sought as necessary. The Uniform Requirements of the International Committee of Medical Journal Editors state that editors must not disclose information about manuscripts even if requested to for legal proceedings.[29] Care must also be taken not to release any confidential information or the names of anyone whose identity must be protected, such as the reviewers. A reviewer may, however, volunteer that his or her identity can be made known. This can sometimes be very helpful and avoid the situation where a reviewer is subsequently contacted by the investigating committee to act as an independent expert witness. No reviewer should, however, be pressured into revealing their identity.

Beware!

Seek legal advice if necessary before referring alleged misconduct to an individual's institution or funding body or taking any action that will bring the alleged misconduct to the notice of others.

Once a case of suspected misconduct has been referred to an outside agency, it is up to that body to impose sanctions if the allegation proves, after investigation, to be founded. Those sanctions may include: a written warning, withdrawal of funding, a request for repayment of grant money, special monitoring of future proposals or work, barring the individual from any further applications for funding, or the implementation of disciplinary procedures. In the medical field, the profession's governing body may in serious cases of misconduct rule that the person can no longer practise. In cases of proven misconduct, a journal may also impose its own sanctions, such as barring the author from submitting to the journal, holding any editorial role or acting as a reviewer. In serious cases of misconduct, criminal proceedings may be brought and the sanctions are then in the hands of the legal system and courts. Publication of an editorial to expose the misconduct is also an option in any case of misconduct, but care must be taken that what is said is within legal restrictions; advice should be sought if there are any doubts or concerns about this.

For work shown to be fraudulent or unreliable, the literature must be corrected and the editor of the journal must make arrangements for this (see this chapter, page 192).

Reviewers found guilty of misconduct

Most editors will want to ban individuals found guilty of reviewing misbehaviour from serving as reviewers again. This may extend to a ban on submitting to the journal or acting in any editorial role. This needs to be dealt with on a case-by-case basis and is very much a personal decision for each editor. A note of caution: it should be remembered that submission bans for any blacklisted author will also, effectively, mean a ban on members of that person's research group or collaborators submitting. These are individuals who have not done anything wrong and who may have joined that person's group or entered into research collaborations with them in ignorance of any misconduct by them, so decisions that will affect them should not be taken lightly. Reporting the individual to his or her institution is also an option in cases of serious misconduct, for example if the reviewer has used privileged information for personal or professional gain.

Beware!

Consider the effect on the innocent members of the group of researchers when imposing sanctions on the individual(s) found guilty of misconduct.

Editors found guilty of misconduct

It is up to each journal, society, owner or publisher to decide what to do with an editor found guilty of misbehaviour. The sanction will depend on the severity of the misconduct. For minor transgressions, a formal letter of displeasure or concern may be appropriate. If the editor's action contravenes his or her contractual obligations, or is actually illegal, he or she may be removed from post. The society or owner may also feel this is necessary if the editor has brought the journal or society into disrepute in some way.

Correcting the literature

The scientific scholarly literature must remain a permanent record of the progress of science – its discoveries and advances, both big and small. It is also a record of the blind alleys, where research hasn't led anywhere or stimulated anyone else to take it any further. That research is there, however, available to be revisited in the future, and perhaps found to be significant in the light of subsequent research and new discoveries. The scientific literature must be protected from corruption, so both genuine errors and data or papers resulting from intended scientific misconduct must be corrected when they are discovered and editors and journals have a duty to

Golden Rule 14

Editors and journals have a duty to keep the scholarly record sound and free from fraudulent or incorrect data.

see that this is done (see Golden Rule 14). Mechanisms exist to do this, and it is the duty of editors to ensure that they act to correct the papers that have been published in their journals when errors are found. Online publication allows a notice of correction to be added any time after an article has been published, and electronic links can be set up between the notice and the paper it refers to, so that both are always viewed together. The published correction can also be linked to citations of the article in online indexing databases.

There are a number of specific terms used for different types of corrections of the literature. They can be confusing, especially as some editors may never have come across them, so a definition of the various terms used may be useful.

Notification of an error

The terms 'erratum' and 'corrigendum' are both used to denote the correction of relatively straightforward and unintended errors. They have distinct meanings but are sometimes confused and used incorrectly. It may be better, therefore, to use just one or other, or the term 'correction'.

An *erratum* is published to correct an error made by the journal or publisher during the publishing process. For example, an error may have been made during copyediting, a figure legend may have been put with the wrong figure, or a bit of text may have been left out by mistake.

A *corrigendum* is published to correct an error made by the authors. For example, a wrong figure may have been sent, the labelling on a figure may be wrong, an author's name may have been misspelt, or even left out, or the wrong materials or cell lines may have been described as being used in the experiments.

Neither errata nor corrigenda should be used for the correction of spellings or language, and other similar cases, unless these have resulted in confusing or ambiguous interpretation. They should be used only to correct errors that affect the content of a paper, that may influence the interpretation of the work or its repetition, or that incorrectly attribute credit for the work. In both cases, the correction should appear on a numbered page in the journal and should be listed on the contents page with the full original citation given. It is important that corrections are always on numbered pages in journals so that they are in citable form and can therefore be readily located. For online journals, a link should be created between the correction and the paper. The United States National Library of Medicine (NLM) provides good real examples of various corrections and how they appear in the NLM (www.nlm.nih.gov/pubs/factsheets/errata.html).

Expression of concern

If investigations into alleged or suspected research misconduct have not yet been completed or prove to be inconclusive, an editor or journal may wish to publish an 'expression of concern', detailing the points of concern and what actions, if any, are in progress. Again, this statement should appear on a numbered page of the journal, be listed on the contents page, the full original citation be given, and a link created between the statement and the original article in the online version of the journal if there is one. A good example can be found in the journal *Science*, where the editor-in-chief, Donald Kennedy, published an 'Editorial expression of concern' to alert readers about the questionable validity of the findings in the two papers the journal had published on the stem-cell work of Woo Suk Hwang as soon as suspicions were raised (see this chapter, page 177).[30]

Retraction of an article

When evidence is found that a paper has a serious problem, for example that it contains fraudulent data or is the result of fraudulent behaviour, contains serious errors in research results, has already been published elsewhere or contains plagiarized material, a 'retraction' is needed. This can be a confusing term because it does not in the great majority of cases mean that an article is physically withdrawn. Removal should actually be avoided unless absolutely necessary (see this chapter, page 196) to maintain the integrity of the scientific record. Rather, a notice should be published in the journal that the work in the article, either in total or in part, can no longer be considered valid or to be an accurate representation. In the days of paper-only journals, it would not have been possible to remove an article as this would have involved physically locating every copy of the article and cutting it out. But in the online-journal world it is, in theory, very easy to remove or alter content after it has been published. It is therefore crucial that editors and publishers understand that this should not be done except in very exceptional circumstances (see this chapter, page 196). Editors and journals are not, however, under any legal obligation to publish retractions. The scientific community therefore relies on their integrity and persistence to keep the literature sound.

Editors need to be aware, however, of the potential legal consequences of publishing retractions. Things are relatively straightforward if all authors agree to a retraction; but if some do not, and challenge the basis on which the decision to retract has been made, this can become problematical.[31] Great care must therefore be taken and appropriate advice sought. Some publishers may indeed, for legal protection, ask that they see and approve all proposed retractions before they are published. There is always the danger that one of the authors might take legal action against the editor, journal or publisher, claiming that his or her reputation has been damaged without just cause. It is therefore very important that the conclusion that a paper is problematical or fraudulent, and the attribution of blame, are based on rigorous and thorough investigation and that this is expressed appropriately.

There are currently no recognized absolute international standards for retraction, but below are some general guidelines that are followed by many publishers.

1   It used to be the case that a retraction had to be signed and agreed by all the authors. This has changed as cases have arisen where not all authors have been prepared to agree to a retraction. Each journal needs, therefore, to decide who it will allow to decide that a retraction is necessary: it may be all or some of the authors, the author's legal representative, the author's institution, or the editor him- or herself. Legal advice may need to be sought before journal policy is decided and implemented.

2   The retraction should appear labelled as such in a prominent section of the journal on a numbered page (so that it is published in citable form) and the full title, author list and citation details of the original article should be given.

3   The retraction should be listed with the original citation details and article title and authors on the contents page of the journal.

4   The text of the retraction should explain why the article is being retracted and it should be made clear whether the whole article is being retracted or just a part, or parts, of it.

5   If the editor has any concerns or comments to make they should be included. But, as with all sensitive issues, care should be taken with what is said and legal advice sought if there is any doubt about what it is legally permissible to say.

6   The retraction statement and the original article must be clearly linked electronically in the online journal so that the retraction will always be apparent to readers. It should ideally precede the article or appear on the same screen so readers are aware of the retraction before they access or read the article itself.

Examples of retractions that were published quickly after the discovery of fraudulent work and that are very clear and informative are those that followed the expression of concern[30] published by the journal *Science* on the two Woo Suk Hwang stem-cell papers (see this chapter, page 194). The retractions were published 1 month later, and included details of the findings of the committee investigating the alleged fraud and how many authors had agreed to the retraction, together with an apology from the journal.[32] An example of the publication of multiple retractions together can be found in the journal *Nature*, where retractions for seven papers authored by Jan Hendrik Schön appeared on two consecutive pages in the journal.[33] Here, the seven papers involved are clearly listed, and, for each, the level of retraction is indicated along with comments from the various authors, including the dissociation of Jan Hendrik Schön from two of the retractions because of his belief in the science presented in those papers. For the other five papers, the authors note that the papers may contain some legitimate ideas and contributions.

Removal of an article

Some editors and publishers may feel that no article should ever be removed from the scientific literature, whatever the circumstances, and that the mechanism of retraction should always be used instead for fraudulent or problematic papers, with appropriately highlighted warnings. Others, however, may consider that in exceptional circumstances article removal is warranted. The International Association of Scientific, Technical and Medical Publishers (STM) provides some guidance on the restricted circumstances when (subject to legal advice) removal is appropriate[34]:

- the privacy of a research subject has been violated in an inappropriate way
- there might be a significant health risk to members of the public if any errors in a publication were followed
- a paper contains defamatory comments about others or their work.

The article should be replaced online with a notice that incorporates the bibliographic metadata and an explanation of why the article has been removed. As for retractions, there are no absolute international standards for article removal. STM believes that scholarly publishers should not in cases of legal infringement (for example where there has been double or multiple publication, and the same article or plagiarized material is published in more than one place) require other publishers to remove articles. It advises that the publisher(s) other than the one with the earliest publication rights publish a retraction note(s) indicating double or multiple publication and referring readers to the article in the journal with the earliest publication rights. Guidance on determining earliest publication rights is provided.

The International Federation of Library Associations and Institutions (IFLA) and the International Publishers' Association (IPA) have issued a joint statement on certain best-practice principles regarding the removal of articles from electronic databases.[35] They have agreed that articles which a publisher has withdrawn, or cannot for legal reasons continue to make available, should be stored in the official archive of the publisher as a record of scholarly transaction and, in certain circumstances, also by libraries. The role of libraries as custodians of the published record is acknowledged and so libraries should, upon request, have and manage access to those articles that are no longer available to the public.

Replacement of an article

Very occasionally, if an article contains errors that might pose a serious health risk, the editor and publisher may decide it should be replaced by a corrected version, following the procedures for retraction but with the notification screen online linked to the corrected article.[36] However, as for article removal, some editors and publishers may feel that no article in the scientific literature should be replaced whatever the circumstances, and prefer that the mechanism of retraction be used instead, with appropriately highlighted warnings.

Dubious or fraudulent data remaining in the literature

Unfortunately, work that has been discredited may continue to be cited and used after it has been shown to be fraudulent or unreliable.[31] This may occur even if the work has been officially retracted in a journal, because a researcher or writer may find the article in a paper copy of the journal and not be aware that it has been retracted. Alternatively, journals may fail to retract an article despite being notified that it should be.[37] Editors and journals have a duty to try to keep the literature as sound and uncontaminated as possible and so should make every attempt to publish whatever corrections are needed when they become aware of problematical or fraudulent data. However, even if they are rigorous in this, it does not guarantee that the problem of continued use of the data will be resolved. In the case of retracted articles, for example, the amount of citation following retraction can be quite significant. Budd *et al.* looked at 235 biomedical articles that had been retracted and found that these were cited 2034 times after the retraction notices.[38] When they looked at a subset of those citations, they found that the great majority of the citations treated the articles as though they represented valid research. Sox and Rennie[37] have also reported on the continuing citation of discredited research after its retraction and have called upon authors who cite articles that are shown to be fraudulent to set the record straight and to publish corrections. They have recommended that journals provide free access to retracted articles and the linked accompanying notices of retraction from the day the retractions are published, stating (page 612) that 'Readers should not have to pay to see the self-correction mechanisms of science at work'. This is something that publishers should perhaps take on board.

The future

From the cases and examples given in this chapter, it will be clear that misconduct appears to be rife. It is unlikely that the problem will disappear, and it is unrealistic to think that there is some magic solution. As long as so much hinges on people's publication records and the potential rewards of some research is so great, the fallibility and failings of humans will continue to surface and misconduct will occur. But this is not to say that nothing can be done to help decrease the incidence of misconduct. Education can play a large part in this, by making individuals realize that certain behaviour is inappropriate, by creating an awareness of ethical issues, and by introducing the concept of good practice in scientific research and publication from an early age. Editors and journals have an important role to play in this educative process.

References

1   Kirby, K. and Houle, F. A. (2004). Ethics and the welfare of the physics profession. *Physics Today*, November, 42. www.physicstoday.org/vol-57/iss-11/p42.html.

2   Martinson, B. C., Anderson, M. S. and de Vries, R. (2005). Scientists behaving badly. *Nature*, **435**, 737–738.

3   Adam, D. and Knight, J. (2002). Journals under pressure: Publish, and be damned. . . . *Nature*, **419**, 772–776.

4   Lock, S. (1985). A Difficult Balance. Editorial peer review in medicine. ISI Press, Philadelphia.

5   Beasley, M. R., Datta, S., Kogelnik, H., Kroemer, H. and Monroe, D. (2002). Report of the Investigation Committee on the Possibility of Scientific Misconduct in the Work of Hendrik Schön and Coauthors. September 2002. www.lucent.com/news_events/pdf/researchreview.pdf (three-page summary available at www.lucent.com/news_events/pdf/summary.pdf).

6   Dalton, R. (2002). Misconduct: The stars who fell to Earth. *Nature*, **420**, 728–729.

7   Hwang, W. S. *et al.* (2004). Evidence of a pluripotent human embryonic stem cell line derived from a cloned blastocyst. *Science*, **303**, 1669–1674.

8   Hwang, W. S. *et al.* (2005). Patient-specific embryonic stem cells derived from human SCNT blastocysts. *Science*, **308**, 1777–1783.

9   Timeline of a controversy. (2005). A chronology of Woo Suk Hwang's stem-cell research. *Nature*, 19 December. doi:10.1038/news051219-3. www.nature.com/news/2005/051219/full/051219-3.html.

10   Giles, J. (2005). Special Report: Taking on the cheats. *Nature*, **435**, 258–259.

11   www.plagiarism.org (accessed 1 February 2006).

12   von Elm, E., Poglia, G., Walder, B. and Tramèr, M. R. (2004). Different patterns of duplicate publication. An analysis of articles used in systematic reviews. *JAMA*, **291**, 974–980.

13   Schein, M. and Paladugu, R. (2001). Redundant surgical publications: Tip of the iceberg? *Surgery*, **129**, 655–661.

14   Pearson, H. (2005). Image manipulation: CSI: cell biology. *Nature*, **434**, 952–953.

15   Rossner, M. and Yamada, K. M. (2004). What's in a picture? The temptation of image manipulation. *Journal of Cell Biology*, **166**, 11–15.

16   Rossner, M. (2006). How to guard against image fraud. *The Scientist*, **20**, 24. www.the-scientist.com/article/display/23156.

17   Office of Research Integrity (ORI). Tools – data imaging. http://ori.dhhs.gov/tools/droplets.shtml (accessed 16 May 2006).

18   North, A. J. (2006). Seeing is believing? A beginners' guide to practical pitfalls in image acquisition. *Journal of Cell Biology*, **172**, 9–18 (online extended bibliography and resources at www.jcb.org/cgi/content/full/jcb.200507103/DC1).

19   Cyranoski, D. (2002). Research materials: Share and share alike? *Nature*, **420**, 602–604.

20   Committee on Publication Ethics (COPE). A code of conduct for editors of biomedical journals. www.publicationethics.org.uk/guidelines/code (accessed 18 February 2006).

21   Committee on Publication Ethics (COPE). COPE 2003 report. www.publicationethics.org.uk/reports/2003/.

22   Savla, U. (2004). When did everyone become so naughty? *Journal of Clinical Investigation*, **113**, 1072.

23   American Society of Plant Biologists. ASPB Ethics in Publishing. www.aspb.org/publications/ethics.cfm (accessed 9 May 2006).

24   Universities UK. (2006). Panel to promote good conduct in medical research. Associated media release, 12 April. www.universitiesuk.ac.uk/mediareleases/riolaunch.asp/.

25   Office of Research Integrity (ORI). (2000). Managing Allegations of Scientific Misconduct: A guidance document for editors. January. http://ori.dhhs.gov/documents/masm_2000.pdf.

26   National Science Foundation. (1997). 'Dear Colleague' letter. 28 January. www.nsf.gov/pubs/1998/oig971/oig971.pdf.

27   Medical Research Council. (1997). MRC Policy and Procedures for Inquiring into Allegations of Scientific Misconduct. Statement by the Medical Research Council. Medical Research Council, London, UK. www.mrc.ac.uk/Utilities/Documentrecord/index.htm?d=MRC002454.

28   Wellcome Trust. Guidelines on Good Research Practice. Including statement on the handling of allegations of research misconduct. www.wellcome.ac.uk (first published January 2002, updated November 2005). Wellcome Trust, London.

29   International Committee of Medical Journal Editors. Uniform Requirements for Manuscripts Submitted to Biomedical Journals: Writing and Editing for Biomedical Publication. www.icmje.org (updated February 2006; accessed 25 May 2006).

30   Kennedy, D. (2005). Editorial expression of concern. *Science*. Published online 22 December. doi:10.1126/science.1124185.

31   Couzin, J. and Unger, K. (2006). Cleaning up the paper trail. *Science*, **312**, 38–43.

32   Kennedy, D. (2006). Editorial retraction of Hwang *et al.*, Science 308(5729) 1777–1783. Retraction of Hwang *et al.*, Science 303(5664) 1669–1674. *Science*, **311**, 335.

33   Retractions of seven papers authored by J. H. Schon. (2003). *Nature*, **422**, 92–93.

34   International Association of Scientific, Technical and Medical Publishers. Preservation of the objective record of science. An STM guideline. www.stm-assoc.org/documents-statements-public-co/ (accessed 5 June 2006).

35   International Federation of Library Associations and Institutions/International Publishers' Association. IFLA/IPA joint statement on removal of articles from databases. www.ipa-uie.org/080705/site%20WEB_REMOV%20OF%20ART.htm (accessed 27 March 2006).

36   ScienceDirect. www.info.sciencedirect.com/licensing/policies/withdrawal/ (accessed 20 March 2006).

37   Sox, H. C. and Rennie, D. (2006). Research misconduct, retraction, and cleansing the medical literature: lessons from the Poehlman case. *Annals of Internal Medicine*, **144**, 609–613.

38   Budd, J. M., Sievert, M. E. and Schultz, T. R. (1998). Phenomena of retraction. Reasons for retraction and citations to the publications. *JAMA*, **280**, 296–297.

Appendix I

The Golden Rules and the Peer-Review Good Practice Checklist

This appendix is made up of two parts: (i) the Golden Rules and (ii) the Peer-Review Good Practice Checklist, which contains the Key Points. The Golden Rules are the most basic and important general principles of peer review, and have been highlighted in the main text; the Key Points have not, but represent a summary of important information that has appeared throughout the book and they form the basis for achieving good practice in peer review. They are grouped under various headings and, in each grouping, guidance is given on what should and what should not be done. The order of the sections follows, roughly, the sequence of considerations and events in peer review.

Not everyone will agree with everything that is in these two sections; and not everyone will agree with the way the information has been divided. Those who are more experienced may think that some of the information in the Key Points is a bit trivial. It may be, but it will very likely be unfamiliar to newcomers to journal editorial work and peer review. As these individuals form a significant proportion of the intended readership, this information is there for their benefit. It may also serve as a refresher for 'old hands', and perhaps act as a catalyst for evaluation and change.

The Golden Rules are not listed in any particular order of importance: they are all important. For that reason, it proved difficult to try to list them that way. So the final order was dictated by the order of appearance in the book.

The Golden Rules

1   Editors are responsible for ensuring the quality of their journals and that what is reported is ethical, accurate and relevant to their readership.

2   Peer review must involve assessment by external reviewers.

3   The submission of a manuscript and all the details associated with it must be kept confidential by the editorial office and all the people involved in the peer-review process.

4   The identity of the reviewers must be kept confidential unless open peer review is used.

5   Reviewers advise and make recommendations; editors make the decisions.

6   Reviewers must assess manuscripts objectively and review the work, not the authors.

7   Editors-in-chief must have full editorial independence.

8   Editorial decisions must be based on the merits of the work submitted and its suitability for the journal; they should not be dictated by commercial reasons, be influenced by the origins of a manuscript, or be determined by the policies of outside agencies.

9   Everyone involved in the peer-review process must always act according to the highest ethical standards.

10 Information received during the submission and peer-review process must not be used by anyone involved for their own or others' advantage or to disadvantage or discredit others.

11 All the parties in the peer-review process must declare any potential conflicts of interest and excuse themselves from involvement with any manuscript they feel they would not be able to handle or review objectively or fairly.

12 No conflict of interest or prejudice must be allowed to influence the submission of a manuscript, its review, or the decision on whether it should be published.

13 Suspected or alleged misconduct must not be ignored.

14 Editors and journals have a duty to keep the scholarly record sound and free from fraudulent or incorrect data.

The Peer-Review Good Practice Checklist

The Key Points

Journal obligations
Journals should:

- make clear their scope, editorial policies, and manuscript presentation and submission requirements
- acknowledge manuscript receipt, record the date of submission, and issue a reference number
- ensure the timely handling and publication of manuscripts submitted to them
- check newly submitted manuscripts to make sure that their content falls within the scope of the journal and that they follow its editorial policy guidelines
- obtain reasons for any requests from authors for changes in authorship or for manuscript withdrawal after submission and ensure these are legitimate and justified
- in optional author-side-pays Open Access models (where authors can pay to make their articles available free online from day of publication), ensure that the

peer-review process and editorial decisions are not influenced by whether or not an author is intending to take up that option.

Journals should not:

- accept a first submission without external review
- compromise reviewer anonymity if closed peer review is used
- get involved in authorship disputes
- get involved in departmental or institutional politics
- allow authors to play off one editor against another in an attempt to force a favourable decision
- make moral or character judgements about authors; actions and decisions on manuscripts should be based solely on the work reported and ethical issues related to it and its submission.

Responsibilities of editors
Editors should:

- ensure their behaviour is transparent and beyond reproach
- develop a written editorial policy and amend and update this regularly to take account of changes in their field and in publishing in general
- ensure manuscripts comply with recognized ethical guidelines and that all procedures at their journals are ethical and in accordance with recommended best practice
- keep manuscript submissions confidential
- ensure that everyone involved in the handling and review of manuscripts understands that they are dealing with privileged information that must not be used for private benefit or gain
- disqualify themselves from handling manuscripts for which a conflict of any kind exists
- ensure the efficient, fair and thorough review of all manuscripts submitted to them and have the appropriate systems in place to achieve this
- request more data or clarification from authors if they come across anything that is unclear or suspicious
- ensure decision making is fair and consistent in their journals
- ensure compliance by authors with their journals' policies, both on submission and after publication; pursue non-compliance and implement appropriate sanctions if it persists
- ensure that peer-reviewed and non-peer-reviewed material is clearly distinguished in their journals
- ensure that any sponsorship of articles is made clear

- keep abreast of developments concerning the publication of research with potentially harmful applications (dual-use research) and introduce appropriate measures and procedures into their journals
- identify manuscripts that report potentially harmful research and ensure that their review is especially rigorous and takes into account the special circumstances
- put in place procedures for dealing with suspected misconduct
- investigate all cases of suspected misconduct at journal level and decide whether there is any substance to the suspicion or claim of misconduct, and then either deal with it themselves or alert the appropriate agency for further investigation and action.

Editors should not:

- abuse the trust of the parties involved in the peer-review process
- personally handle manuscripts from their own institutions or their own research groups
- deliberately choose reviewers who will provide either a favourable or an unfavourable review, or who will hold up the review of a manuscript because they are known to be slow
- use privileged information for personal gain or to disadvantage or discredit others
- attempt to increase the Impact Factors of their journals by unethical means during the peer-review process, for example by inappropriately requesting additional citations to their own journals or deleting citations to competing journals.

Author submission
Authors should:

- choose the most appropriate journal to which to submit their work
- decide which individual will act as corresponding author and give that person responsibility for co-ordinating all issues related to submission and review, including ensuring that all authorship disagreements are resolved appropriately
- submit original work that has been honestly carried out according to rigorous experimental standards
- always give credit to the work and ideas of others that led to their work or influenced it in some way
- declare all sources of research funding and support
- submit manuscripts that are within the scope of journals, ensure that they abide by all those journals' policies and follow all their presentation and submission requirements
- explain in a cover letter if there are any special circumstances, if their manuscript deviates in any way from a journal's requirements or if anything is missing

- ensure that their manuscripts do not contain plagiarized material or anything that is libellous, defamatory, indecent, obscene or otherwise unlawful, and that nothing infringes the rights of others
- ensure they have permission from others to cite personal communications from them and that the extent, content and context have been approved by those individuals
- provide details of related manuscripts they have submitted or have in press elsewhere
- check the references they cite carefully to ensure the details are correct
- notify a journal if work done subsequent to the submission of their manuscript casts doubt on the work submitted or alters its interpretation
- if they decide to submit to another journal after an unsuccessful submission, reformat the manuscript to meet the requirements of the new journal and redraft the cover letter before resubmitting the manuscript.

Authors should not:

- be influenced by the sponsors of their research regarding the analysis and interpretation of their data or in their decision on what to, or not to, publish and when to publish
- divide up their papers inappropriately into smaller ones (minimum publishable units or MPUs) in an attempt to increase their list of publications
- be involved in 'ghost' or 'gift' authorship
- submit the same or a very similar manuscript to more than one journal at the same time
- present their work, or use language, in a way that detracts from the work or ideas of others
- use information obtained privately without direct permission from the individuals from whom it has been obtained
- make exaggerated claims about the novelty or significance of their findings
- misrepresent or inappropriately enhance their results by any means
- make significant changes to their manuscript after acceptance without the approval of the editor or journal editorial office
- submit a manuscript that has been rejected by one journal to another journal without the reviewers' comments being considered and appropriate revisions being made and presentational errors corrected.

Managing the review process
Journals should:

- always treat reviewers with courtesy and respect
- send manuscripts to reviewers, or give instructions on how to access them, as soon as possible after they have agreed to review them

- send reviewers manuscripts that are correctly formatted, well presented and complete, with all ancillary materials included
- provide reviewers with clear instructions and guidance on the journal's aims and scope and what is expected of them in the review process
- instruct reviewers that their narrative reports for authors must correspond to what they have indicated in their confidential reviewing forms and checklists
- give reviewers access to any closely related manuscripts by the authors that are in press or submitted elsewhere for publication
- also have supplementary material that is to be published with a paper peer reviewed as it is an integral part of the publication
- alert reviewers to the possibility that they may be identified if they access material directly on an author's website; they should make alternative arrangements for reviewers to access or receive that material if it is important for the review of a manuscript
- provide reviewers with contact details they can use if they have problems or need assistance during the review of a manuscript
- answer reviewers' queries promptly and sort out any problems as quickly as possible
- ensure that reminder messages sent to reviewers are always courteous and never aggressive or threatening
- thank reviewers for their efforts and give them feedback on the outcome of the review process
- halt the review process if misconduct by the authors is suspected
- remove from the review of manuscripts any reviewers or editors who have acted inappropriately.

Journals should not:

- send to external reviewers manuscripts that are out of scope or do not follow essential journal editorial policy
- send out for review manuscripts in which the standard of language is very poor
- compromise reviewer anonymity in closed peer-review systems
- bombard reviewers with inappropriate review reminders
- automatically send manuscripts that receive opposing opinions or recommendations out for further review.

Reviewer selection
Journals should:

- have a database of reviewers and ensure this is kept up to date
- recognize that reviewer selection is the most critical aspect of peer review
- advise authors on who it is not permissible for them to suggest as potential reviewers for their manuscripts

- ask authors to provide reasons for any requests for exclusion of particular reviewers for their manuscripts
- monitor reviewers' workloads to ensure they are not overloaded or taken advantage of
- have reviewer-selection procedures that involve active decision making
- keep a record, or audit trail, of where various reviewer suggestions have come from
- contact potential reviewers and obtain their agreement to review before they are sent a manuscript
- ensure reviewers are sent manuscripts that are appropriate to their areas of interest and expertise.

Journals should not:

- send manuscripts to more reviewers than are needed with the intention of using only the first reviews returned
- put pressure on any reviewer who feels uncomfortable about reviewing a manuscript to do so
- send manuscripts to reviewers who regularly fail to return reviews, or who do so only after unacceptably long times, or who provide superficial or inadequate reviews.

Reviewer behaviour
Reviewers should:

- provide timely reviews that are both relevant and constructive
- declare any conflicts of interest, either real or potential
- disqualify themselves from review if they feel unable, for any reason, to provide an honest and unbiased assessment
- notify journals of any limitations to their ability to review a manuscript
- declare if they have reviewed a manuscript previously for another journal; if both the editor and reviewer agree that the reviewer can be involved in a second review, he or she should review the manuscript afresh and submit a review based on that assessment
- declare a conflict if asked to review a manuscript that is very similar to one they have submitted elsewhere or have in preparation
- keep confidential the submission and contents of manuscripts sent to them for review
- alert the editor or journal if any circumstances arise that will delay their review
- provide comments that can be forwarded to the author separately from any confidential comments for the editor
- make sure that their comments for authors correspond to their assessment on the confidential review forms and checklists

- report to journals any suspicions of misconduct and ask for advice on how to proceed.

Reviewers should not:

- agree to review a manuscript just to gain sight of it for personal benefit with no intention of providing a genuine review
- contact anyone else about reviewing a manuscript without the knowledge and permission of the journal from which it was received
- use information in manuscripts they review for their own or others' benefit or gain, or plagiarize any of the material within those manuscripts
- intentionally delay return of their reports
- make personal or derogatory comments about authors in their reviews
- request that authors include citations to their own work in order to receive additional citations for themselves
- contact the authors directly about any manuscript they review.

Handling reviews
Journals should:

- check reviews as they are submitted rather than when all the reviews are in, so that errors can be corrected, clarification obtained, and additional action taken if necessary
- ensure that reviewers' reports for authors do not contain anything that is defamatory, libellous or likely to confuse the authors
- check before editorial decisions are made that the correct reviews have been assessed and that all the reviews submitted have been considered
- recognize that the most important part of a review is that which contains the detailed comments, not the recommendation.

Journals should not:

- selectively edit reviewers' reports so that they better reflect an editorial decision.

Editorial decision-making
Editorial decisions should:

- be made or approved by an editor, and should not default to someone who does not have appropriate specialist training or knowledge
- be based on the merits of the work submitted and its suitability for the journal
- give more weight to reports of flaws or technical criticisms than to subjective opinions on suitability for a journal
- be consistent across all manuscripts submitted to a journal
- be as transparent as possible; editors should be able to substantiate their decisions if challenged

- provide reasons for any comments and opinions in reviewers' reports that have been overridden by the editor.

Editorial decisions should not:

- be influenced by the origins of manuscripts or determined by the policies of outside agencies
- except in exceptional circumstances (for example, misconduct or if a serious flaw comes to light), override all the reviewers' recommendations and opinions
- allow papers to be published with overstated claims or interpretations.

Feedback to authors
Journals should:

- notify authors if the review process is unduly extended or additional review is required
- if a manuscript is withdrawn from review by the journal, give the authors clear reasons why this is being done and provide them with the opportunity to re-spond if appropriate
- check decision letters before these are communicated to authors to ensure that editorial notes, instructions, and inappropriate words or phrases have been removed
- make clear to authors exactly what the decision is on their manuscript, the rea-sons for it and, if appropriate, what conditions need to be met for the journal to consider the manuscript again
- let authors know what the procedures will be for handling a resubmission of their manuscript
- answer authors' queries promptly and informatively
- keep a full record of all manuscript status enquiries to ensure accuracy and consistency in responses.

Journals should not:

- misinform authors about the review status of their manuscript
- allow editorial staff to tell authors informally the decisions on their manuscripts based on editors' initial recommendations; this should not be done until deci-sions have been finalized.

Revised manuscripts
Journals should:

- remind authors of revision deadlines
- accept a manuscript if an author has fulfilled all the revision and journal policy requirements within the stipulated time
- send authors official, dated acceptance letters.

Journals should not:

- issue a blanket instruction that all revised manuscripts are to be sent to the original reviewers for assessment; decisions on review procedure should be made on a case-by-case basis.

Accepted manuscripts
Journals should:

- check manuscripts before moving them on for preparation for publication to ensure they are complete and all the required information and enclosures have been received
- refer for editorial approval requests from authors both for non-trivial changes to a manuscript after it has been accepted and for notes to be added in proof
- publish papers in a timely and efficient manner, with dates of receipt and acceptance.

Journals should not:

- reverse the decision to accept a paper for publication unless a serious problem is subsequently found, for example fraud or an ethical issue; such decisions should not be reversed because a journal has misjudged the availability of space.

Authors should:

- supply any missing items or information promptly when requested by journals
- abide by all the post-publication policy requirements of journals
- notify a journal immediately if errors are found in a paper after publication so that an appropriate correction note can be published if necessary.

Authors should not:

- include in notes added in proof information they were aware of and should have either included or referenced in their manuscript because it was already published at the time of submission.

Dealing with misconduct
Journals should:

- have procedures in place for investigating allegations of misconduct at journal level
- look into all cases of suspected misconduct
- take extra care to keep complete and accurate records when suspicions or allegations of misconduct arise
- give the person(s) against whom allegations have been made the opportunity to respond to the allegations and to provide an explanation

- obtain written evidence from all the parties involved
- refer to individuals' institutions or funding bodies cases that warrant further investigation
- ensure that referral to an individual's institution or funding body is warranted as this is a serious step with potentially very serious implications for the individual and his or her reputation
- reprimand individuals found guilty of misconduct or inappropriate behaviour and implement appropriate sanctions
- publish appropriate correction notes for papers they have published in which errors, fraudulent data or misconduct have been found
- ensure that when correcting the literature any conclusion that a paper is problematical or fraudulent and the attribution of blame are based on rigorous and thorough investigation and expressed appropriately and within legal restrictions.

Journals should not:

- mistake genuine errors for misconduct
- launch full-scale external investigations into allegations of research misconduct to determine if they are substantiated and that misconduct has occurred; this is the responsibility of employers and funding agencies
- release information about allegations or suspicions of suspected misconduct until they have been substantiated
- pressure reviewers into revealing their identities to investigating bodies in misconduct cases
- alter papers or remove them from the scholarly literature once they have been published except in very exceptional and restricted circumstances.

Appendix II

Examples of checklists, forms, guidance for reviewers and editorial letters

This appendix gives examples of various documents that editorial offices need to use – checklists, forms, guidance notes and editorial letters. There are no global unique versions that will suit all journals. Journals have different requirements and so need to adapt and tailor documents to suit both the way they work and the characteristics and needs of the communities they serve. But it is often helpful to see what other journals are using and what they consider is necessary. Having good and appropriate documents makes manuscript handling easier, and can greatly increase the efficiency and quality of review. But don't just forget about these documents once you've introduced or initially revised them when taking on a new journal or editorial role. They should be revisited periodically and updated to make sure they are still necessary and appropriate, and new ones need to be drafted to deal with changing editorial or policy requirements.

Some of the items – such as the checklists – are general basic composites I have put together. Other items, such as the conflict-of-interest forms and reviewing forms and guidance notes, are 'real' and the versions in use in August 2006 are reproduced with the kind permission of various journals and organizations. These items were selected because they are particularly good or helpful, or have something that isn't used by the majority of journals but which some might want to consider. In the latter case, there is sometimes a range of examples (such as for the conflict-of-interest forms), suitable for different levels of need. Website addresses are given for the real documents so readers can view the latest versions.

(1) Checklists

(a) Checklist for new manuscript submissions

New Manuscript Checklist				
Manuscript ref. no: **Corresponding author:**			Date rec'd	Editor

Manuscript type: Article   Review   Letter   Editorial

Resubmission: No   Yes ⟶   – previous ref. no ....
   – response letter   No   Yes

Supplementary material: No   Yes ⟶   – for external review   No   Yes
   – to be published   No   Yes
   – file size OK   No   Yes
   – any author websites   No   Yes

Colour: No   Yes ⟶   – figures ....
   – colour fee waiver requested   No   Yes

Cover image: No   Yes ⟶   – where located ....

Related manuscript(s):   No   Yes ⟶   – for external review   No   Yes

Pages numbered:	No Yes		**Word count:**	....	
Section order OK:	No Yes		**No. tables:**	....	
Title OK:	No Yes		**No. figures:**	....	
Running head OK:	No Yes		**Figure quality OK:**	No Yes	
Abstract OK:	No Yes		**All legends included:**	No Yes	
Keywords:	No Yes		**English OK:**	No Yes	

Author list checked:	No Yes	**All affiliations given:**	No Yes
Suggested reviewers:	No Yes	**Non-preferred reviewers:**	No Yes

Any competing interests: No   Yes ⟶   what? ....

Any policy restrictions: No   Yes ⟶   what? ....

Anything missing: No   Yes ⟶   what? ....

Any problems: No   Yes ⟶   what? ....

Any special requests: No   Yes ⟶   what? ....

	Date to editor	Notes
Handled by: **Date complete/ready for transfer:**		

(b) Checklist for revised manuscripts

Revised Manuscript Checklist		

Manuscript ref. no: Corresponding author:	Date rec'd	Editor

Manuscript type: Article   Review   Letter   Editorial

Supplementary material:	No Yes ⟶	– to be published	No Yes
		– file size OK	No Yes
Colour:	No Yes ⟶	– figures	
		– colour fee waiver	No Yes
		– payment form rec'd	No Yes
Cover image:	No Yes ⟶	– where located	

Response letter for: – editor comments   No   Yes   – reviewer comments   No   Yes

Marked-up copy MS:	No Yes
'Clean' copy MS:	No Yes
All forms rec'd:	No Yes

Pages numbered:	No Yes	**Word count:**	....
Section order OK:	No Yes	**No. tables:**	....
Title OK:	No Yes	**No. figures:**	....
Title changed:	No Yes	**Figure quality OK:**	No Yes
Running head OK:	No Yes	**Figure format OK:**	No Yes
Abstract OK:	No Yes	**All legends included:**	No Yes
Keywords:	No Yes	**English OK:**	No Yes

Author list checked: No Yes	**All affiliations given:** No Yes

Any author changes:	No Yes ⟶ what?/reasons ....
Any competing interests:	No Yes ⟶ what? ....
Any policy restrictions:	No Yes ⟶ what? ....
Any permissions outstanding: No Yes ⟶ what?	

Anything missing: No   Yes ⟶   what? ....
Any problems: No   Yes ⟶   what? ....
Any special requests: No   Yes ⟶   what? ....

Handled by: **Date complete/ready for assessment:**	Date to editor	Notes

(c) Checklist for sending accepted manuscripts on to production

Accepted Manuscript Checklist – for production	
Manuscript ref. no:	Date rec'd:
Corresponding author:	Date revised:
CA's email:	Date accepted:
Any dates CA will be away/unavailable:	Date to production:
Alternative contact:	Production no:
	Issue:

Manuscript type: Article   Review   Letter   Editorial

Supplementary material: No   Yes   →   – what? ....
                                          – no. files ....
                                          – file sizes ....
                                          – notes ....

Colour: No   Yes   →   – figures ....

Colour payment form: No   Yes   →   – notes ....

Colour fee waiver: No   Yes   →   – reason granted ....

Exclusive licence form: No   Yes   →   – notes ....

Open access option: No   Yes   →   – notes ....

To be pub. back-to-back: No   Yes   →   – with MS ....
                                          – MS order ....

No. tables: ....

No. figures: ....

Better figures req'd: No   Yes   →   – which? ....

Level of copyedit req'd: ....

Any nomenclature issues: No   Yes   →   – what? ....

Items still to be supplied by Au: ....

Notes:

(d) CONSORT checklist and flowchart

The CONSORT statement (www.consort-statement.org) is described in Appendix III (see page 270). It comprises a checklist and a flowchart, and these are reproduced as Examples 1 and 2 with kind permission from the CONSORT Group. The current versions can be downloaded from www.consort-statement.org/Downloads/download.htm and further information on the use of CONSORT can be found in the article by D. Moher *et al.* which was published simultaneously in three journals in 2001 ('The CONSORT statement: revised recommendations for improving the quality of reports of parallel-group randomized trials'; *Annals of Internal Medicine*, **134**, 657–662; *JAMA*, **285**, 1987–1991; and *The Lancet*, **357**, 1191–1194). The statement is due to be revised in January 2007.

CONSORT Checklist of items to include when reporting a randomized trial

PAPER SECTION And topic	Item	Description	Reported on Page #
TITLE & ABSTRACT	1	How participants were allocated to interventions (*e.g.*, "random allocation", "randomized", or "randomly assigned").	
INTRODUCTION Background	2	Scientific background and explanation of rationale.	
METHODS Participants	3	Eligibility criteria for participants and the settings and locations where the data were collected.	
Interventions	4	Precise details of the interventions intended for each group and how and when they were actually administered.	
Objectives	5	Specific objectives and hypotheses.	
Outcomes	6	Clearly defined primary and secondary outcome measures and, when applicable, any methods used to enhance the quality of measurements (*e.g.*, multiple observations, training of assessors).	
Sample size	7	How sample size was determined and, when applicable, explanation of any interim analyses and stopping rules.	
Randomization -- Sequence generation	8	Method used to generate the random allocation sequence, including details of any restrictions (*e.g.*, blocking, stratification)	
Randomization -- Allocation concealment	9	Method used to implement the random allocation sequence (*e.g.*, numbered containers or central telephone), clarifying whether the sequence was concealed until interventions were assigned.	
Randomization -- Implementation	10	Who generated the allocation sequence, who enrolled participants, and who assigned participants to their groups.	
Blinding (masking)	11	Whether or not participants, those administering the interventions, and those assessing the outcomes were blinded to group assignment. When relevant, how the success of blinding was evaluated.	
Statistical methods	12	Statistical methods used to compare groups for primary outcome(s); Methods for additional analyses, such as subgroup analyses and adjusted analyses.	
RESULTS Participant flow	13	Flow of participants through each stage (a diagram is strongly recommended). Specifically, for each group report the numbers of participants randomly assigned, receiving intended treatment, completing the study protocol, and analyzed for the primary outcome. Describe protocol deviations from study as planned, together with reasons.	
Recruitment	14	Dates defining the periods of recruitment and follow-up.	
Baseline data	15	Baseline demographic and clinical characteristics of each group.	
Numbers analyzed	16	Number of participants (denominator) in each group included in each analysis and whether the analysis was by "intention-to-treat". State the results in absolute numbers when feasible (*e.g.*, 10/20, not 50%).	
Outcomes and estimation	17	For each primary and secondary outcome, a summary of results for each group, and the estimated effect size and its precision (*e.g.*, 95% confidence interval).	
Ancillary analyses	18	Address multiplicity by reporting any other analyses performed, including subgroup analyses and adjusted analyses, indicating those pre-specified and those exploratory.	
Adverse events	19	All important adverse events or side effects in each intervention group.	
DISCUSSION Interpretation	20	Interpretation of the results, taking into account study hypotheses, sources of potential bias or imprecision and the dangers associated with multiplicity of analyses and outcomes.	
Generalizability	21	Generalizability (external validity) of the trial findings.	
Overall evidence	22	General interpretation of the results in the context of current evidence.	

Example 1 CONSORT checklist (reproduced with permission from the CONSORT Group).

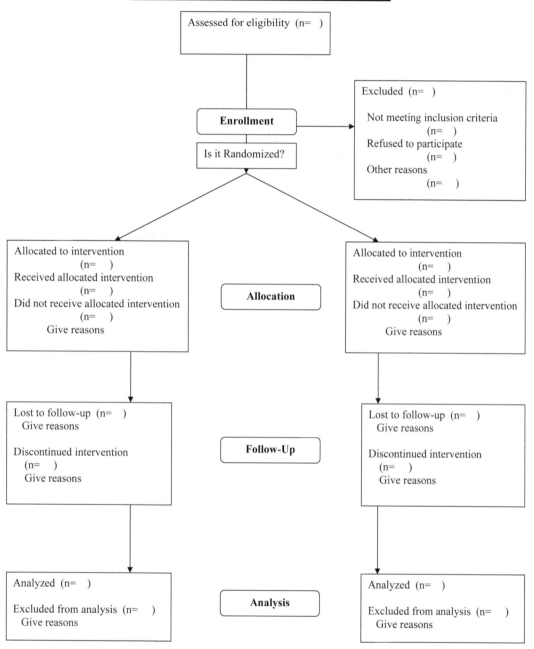

Example 2 CONSORT flowchart (reproduced with permission from the CONSORT Group).

(2) Forms

(a) Change of authorship form

The 'Change of authorship form' reproduced as Example 3 is used with kind permission from the American Physiological Society (www.the-aps.org). It can be accessed from www.the-aps.org/publications/i4a/revision.htm and a PDF of the form can be found at www.the-aps.org/publications/journals/pub_author_form.PDF.

CHANGE OF AUTHORSHIP FORM
Must be completed and signed by ALL authors

Please check all that apply
- New author(s) have been added.
- There is a change in the order of authorship.
- An author wishes to remove his/her name. An author's name may only be removed at his/her own request and a letter signed by the author should accompany this form.

Manuscript Number _____

Manuscript Title _____

Former Authorship
Please list **ALL AUTHORS** *in the same order* as the original submission. For more than 12 use an extra sheet.

	Print Name		Print Name
Name (1)		Name (7)	
Name (2)		Name (8)	
Name (3)		Name (9)	
Name (4)		Name (10)	
Name (5)		Name (11)	
Name (6)		Name (12)	

Copyright Transfer. In consideration of the acceptance of the above work for publication, I do hereby assign and transfer to the American Physiological Society (APS) all rights, title, and interest in and to the copyright in the above-titled work. This includes preliminary display/posting of the abstract of the accepted article in electronic form before publication. In cases in which obtaining a signature from each author would delay review, the corresponding author's signature is sufficient provided that the corresponding author understands that he or she signs on behalf of all of the authors who have not signed the form. It should also be understood by all authors that the corresponding author has signed the manuscript submission form as their proxy. If any changes in authorship (order, deletions, or additions) occur after the manuscript is submitted, agreement by all authors for such changes must be on file with APS. An author's name may only be removed at his/her own request. For commercial companies, authorized agents' signatures are allowed for copyright transfer, but the corresponding author must sign for authorship responsibilities as stated below. Material prepared by employees of the US Government in the course of their official duties cannot be copyrighted.

Authorship Responsibilities. I attest that:
1) the manuscript is not currently under consideration, in press, or published elsewhere, and the research reported will not be submitted for publication elsewhere until a final decision has been made as to its acceptability by the journal **(posting of submitted material on a web site may be considered prior publication--note this in your cover letter)**;
2) the manuscript is truthful original work without fabrication, fraud, or plagiarism;
3) I have made an important scientific contribution to the study and am thoroughly familiar with the primary data; and
4) I have read the complete manuscript and take responsibility for the content and completeness of the manuscript and understand that I share responsibility if the paper, or part of the paper, is found to be faulty or fraudulent.

☐ **Please check if this article was written as part of the official duties of an employee of the US Government**.

Conflict of Interest Disclosure. All funding sources supporting the work and all institutional or corporate affiliations of mine are acknowledged. Except as disclosed on a separate attachment, I certify that I have no commercial associations (e.g., consultancies, stock ownership, equity interests, patent-licensing arrangements) that might pose a conflict of interest in connection with the submitted article, and that I accept full responsibility for the conduct of the trial, had full access to all the data, and controlled the decision to publish.
☐ **I have a potential conflict to disclose and will provide the form from** http://www.the-aps.org/publications/journals/conflict.pdf.

NEW AUTHORSHIP - All authors must sign below agreeing to the new changes in authorship. The authorship order **must** match the new title page of the manuscript.

Name (1)		Signature		Date	
Name (2)		Signature		Date	
Name (3)		Signature		Date	
Name (4)		Signature		Date	
Name (5)		Signature		Date	
Name (6)		Signature		Date	
Name (7)		Signature		Date	
Name (8)		Signature		Date	
Name (9)		Signature		Date	
Name (10)		Signature		Date	
Name (11)		Signature		Date	
Name (12)		Signature		Date	

Please fax to 301-634-7243

Example 3 Change of authorship form (used with permission from the American Physiological Society).

(b) Conflict-of-interest forms

A number of examples of forms requiring individuals to declare potential conflicting or competing interests are given below. They vary in the level of detail requested, from comprehensive and fairly prescriptive (as for journals (i)–(iii)), to ones where the onus is placed on the individuals to describe potential competing interests in their own statements (as for journal (iv)). Most conflict-of-interest forms are aimed at authors, but some journals also ask their reviewers to make declarations (for example, as does journal (iii)). Some journals have also decided that readers should know of any competing interests their editors may have, and so publish these for each of their associate editors (as does journal (v)).

(i) The journal *Nature* (www.nature.com) has a 'Competing financial interests declaration form' that authors are required to return before final acceptance of their contributions. Definitions of what are considered competing interests and guidance on use of the form can be found at www.nature.com/nature/authors/policy/competing.html. The actual form can be downloaded from www.nature.com/nature/authors/policy/form.html and is reproduced as Example 4 with kind permission from *Nature*.

Competing financial interests declaration form

In the interests of transparency and to help readers to detect potential bias, we now require authors of primary and secondary research papers to declare any competing financial interests in relation to papers accepted for publication .

Please submit one statement on behalf of all authors.

Competing financial interests are defined as those that, through their potential influence on behaviour or content or from perception of such potential influences, could undermine the objectivity, integrity or perceived value of a publication. They may include any of the following:

Funding: Support for a research programme (including salaries, equipment, supplies, reimbursement for attending symposia, and other expenses) by organizations that may gain or lose financially through publication of this paper.

Employment: Recent (i.e. while engaged in this research project), present or anticipated employment by any organization that may gain or lose financially through publication of this paper.

Personal financial interests: Stocks or shares in companies that may gain or lose financially through publication; consultation fees or other forms of remuneration from organizations that may gain or lose financially; patents or patent applications whose value may be affected by publication.

It is difficult to specify a threshold at which a financial interest becomes significant, although we note that many US universities require faculty members to disclose interests exceeding $10,000 or 5% equity in a company (see for example Lo, B. *et al., New Eng. J. Med.* **343,** 1616). Any such figure is necessarily arbitrary, however, so we offer as one possible practical alternative guideline: "Any undeclared competing financial interests that could embarrass you were they to become publicly known after your work was published."

We do not consider diversified mutual funds or investment trusts to constitute a competing financial interest.

We do not require authors to state the monetary value of their financial interests.

Please print, complete and return the following form.

Example 4 Form for declaration of competing financial interests by authors (reproduced with permission from *Nature*).

Declaration of competing financial interests

JOURNAL:

MANUSCRIPT TITLE:

MANUSCRIPT NUMBER:

Please check one of the following:

1.   ⌐   I declare that the authors have no competing interests such as those defined above or others that might be perceived to influence the results and discussion reported in this paper.

2.   ⌐   The authors have competing interests such as those defined above or others that might be perceived to influence the results and discussion reported in this paper.

3.   ⌐   I decline to respond to this request for information.

If you have checked 2, please specify the competing interests:

In reports of research, under a heading "Competing interests", we will publish "Authors declare competing financial interests: see Web version for details", "Authors declare they have no competing financial interests" or "Authors decline to provide information about competing financial interests"

SIGNATURE:

NAME IN BLOCK CAPITALS:

DATE:

Example 4 (*Cont'd*)

(ii) The journal *Science* (www.sciencemag.org) has a 'Statement on real or perceived conflicts of interest for authors' and all authors must, when uploading their revised manuscripts after peer review, fill in details of financial and other interests that might pose a conflict of interest. The statement and form can be found at www.sciencemag.org/feature/contribinfo/prep/coi.pdf and are reproduced as Example 5 with kind permission from *Science*.

Statement on Real or Perceived Conflicts of Interest for Authors

Information pertaining to *all* authors must be entered on this form

Science has a primary responsibility to its readers and to the public to provide in its pages clear and unbiased scientific results and analyses. Although we rely on the expertise of our Board of Reviewing Editors and our peer reviewers to help us accomplish this, we think that our readers should be informed of additional relationships of our authors that could pose a conflict of interest. Thus, for readers to evaluate the data and opinions presented in *Science*, they must be informed of financial and other interests of our authors that may be at odds with unbiased presentation of data or analysis.

Therefore, *Science* believes that manuscripts (Brevia, Essays, Perspectives, Policy Forums, Reports, Research Articles, Reviews, and Viewpoints) should be accompanied by clear disclosures from all authors of their affiliations, funding sources, or financial holdings that might raise questions about possible sources of bias. Disclosure is accomplished in three ways:

First, by a complete listing of the current institutional affiliations of the authors. This list must include academic as well as corporate and other industrial affiliations. As the editors deem appropriate, items in this list will be included in the author affiliations printed in the manuscript. Please indicate below:

❑   All affiliations of all authors are listed on the title page of the paper.
❑   Additional affiliations not on the title page are:

Second, through the acknowledgment of all financial contributions to the work being reported, including contributions "in kind."   All funding sources will be listed in the published manuscript. Please indicate below:

❑   All funding sources for this study are listed in the acknowledgement section of the paper.
❑   Additional funding sources not noted in the manuscript are:

Example 5 Form for statement of real or perceived conflicts of interest by authors (reproduced with permission from *Science*).

Third, through the execution of a statement disclosing to the Editors all financial holdings, professional affiliations, advisory positions, board memberships, patent holdings and the like that might bear a relationship to the subject matter of the contribution. The Editors will determine whether the material disclosed to them should be published as part of the article. Please check the appropriate box below:

The following are declarable relationships:

Financial: Significant financial interest (equity holdings or stock options) in any corporate entity dealing with the material or the subject matter of this contribution. Please disclose the entity and the nature and amount of the holding.
❑ None
❑ One or more authors has a financial relationship, as described below.

Management/Advisory affiliations: Within the last 3 years, status as an officer, a member of the Board, or a member of an Advisory Committee of any entity engaged in activity related to the subject matter of this contribution. Please disclose the nature of these relationships and the financial arrangements.
❑ None
❑ One or more authors has a management/advisory relationship, as described below

Paid Consulting: Within the last 3 years, receipt of consulting fees, honoraria, speaking fees, or expert testimony fees from entities that have a financial interest in the results and materials of this study. Please enumerate.
❑ None
❑ One or more authors has a paid consulting relationship, as described below

Patents: A planned, pending, or awarded patent on this work by any of the authors or their institutions. Please explain.
❑ None
❑ One or more authors or the authors' institutions has a patent related to this work, as described below

❑ All authors declare that we have read Science's full Conflict of Interest Policy and have disclosed all declarable relationships as defined therein, if any.

Manuscript Number_____

Title:_____

_____

First Author:_____

Signature:_____ Date:_____

Example 5 (*Cont'd*)

(iii) The *BMJ* (*British Medical Journal*; http://bmj.bmjjournals.com) asks both authors and reviewers to declare all competing financial interests and suggests that individuals might also want to disclose any other potential competing interests. The 'Declaration of competing interest' forms, and guidance on filling them in, can be found at http://bmj.bmjjournals.com/cgi/content/full/317/7154/291/DC1, and are reproduced as Example 6 with kind permission from the *BMJ*.

BMJ DECLARATION OF COMPETING INTEREST

Guidance for authors

To be completed by all authors before publication can go ahead   Article No _____

A competing interest exists when professional judgment concerning a primary interest (such as patients' welfare or the validity of research) may be influenced by a secondary interest (such as financial gain or personal rivalry). It may arise for the authors of a BMJ article when they have a financial interest that may influence -- probably without their knowing -- their interpretation of their results or those of others.

We, the editors of the BMJ, believe that to make the best decision on how to deal with a paper we should know about any such competing interest that authors may have. We are not aiming at eradicating competing interests -- they are almost inevitable. We will not reject papers simply because you have a competing interest, but we will make a declaration on whether or not you have competing interests.

We used to ask authors about any competing interests, but we have decided to restrict our request to financial interests. This is largely a tactical move. We hope that it will increase the number of authors who disclose competing interests. Our experience, supported by some research data, was that authors often did not disclose them.

Please answer the following questions (**all authors must answer**)

1. Have you in the past five years accepted the following from an organisation that may in any way gain or lose financially from the results of your study or the conclusions of your review, editorial, or letter:

_____ Reimbursement for attending a symposium?

_____ A fee for speaking?

_____ A fee for organising education?

_____ Funds for research?

_____ Funds for a member of staff?

_____ Fees for consulting?

Example 6 Forms for declaration of competing interests by authors and reviewers (reproduced with permission from the *BMJ*).

2. Have you in the past five years been employed by an organisation that may in any way gain or lose financially from the results of your study or the conclusions of your review, editorial, or letter?

3. Do you hold any stocks or shares in an organisation that may in any way gain or lose financially from the results of your study or the conclusions of your review, editorial, or letter?

4. Have you acted as an expert witness on the subject of your study, review, editorial, or letter?

5. Do you have any other competing financial interests? If so, please specify.

If you have answered "yes" to any of the above 5 questions, we consider that you may have a competing interest, which, in the spirit of openness, should be declared. Please draft a statement to publish with the article. It might, for example, read:

Competing interests: RS has been reimbursed by Shangri La Products, the manufacturer of elysium, for attending several conferences; TD has been paid by Shangri La Products for running educational programmes and has her research registrar paid for by the company; JS has shares in the company.

If you did not answer "yes" to any of the four questions above, we will publish "Competing interests: None declared." (But see next paragraph)

We are restricting ourselves to asking directly about competing financial interests, but you might want to disclose another sort of competing interest that would embarrass you if it became generally known after publication. The following list gives some examples.

(a) A close relationship with, or a strong antipathy to, a person whose interests may be affected by publication of your paper.

(b) An academic link or rivalry with somebody whose interests may be affected by publication of your paper.

(c) Membership of a political party or special interest group whose interests may be affected by publication of your paper.

(d) A deep personal or religious conviction that may have affected what you wrote and that readers should be aware of when reading your paper.

If you want to declare such a competing interest then please add it to your statement.

To learn more about the thinking that has led to this policy please read the editorial by Richard Smith BMJ 1998;317:291-2.

Example 6 (*Cont'd*)

Please complete option 1 or 2 as appropriate and sign below. If you answered "yes" to any of the 5 questions relating to financial competing interests (or you wish to disclose a non-financial competing interest), you should write a statement below.

It is important that you return this form as early as possible in the publication process. We will not publish your article without completion and return of the form.

☐ 1. Please insert "None declared" under competing interests

or

☐ 2. Please insert the following statement under competing interests:

Title of paper:

Date:

      Signature (all authors to sign):   *(Print name too please)*

Example 6 *(Cont'd)*

Guidance for referees

A competing interest exists when professional judgment concerning a primary interest (such as patients' welfare or the validity of research) may be influenced by a secondary interest (such as financial gain or personal rivalry). It may arise for the referees of a *BMJ* article when they have a financial interest that may influence-probably without their knowing-their interpretation of an article.

We, the editors of the *BMJ*, believe that to make the best decision on how to deal with a paper we should know about any such competing interests that referees may have. We are not aiming at eradicating competing interests-they are almost inevitable. We will not reject opinions simply because you have a competing interest, but we would like to know about it.

We used to ask authors and referees about any competing interests, but we have decided to restrict our request to financial interests. This is largely a tactical move. We hope that it will increase the number of people who disclose competing interests. Our experience, supported by some research data, was that people often did not disclose them.

Please answer the following questions

1. Have you in the past five years accepted the following from an organisation that may in any way gain or lose financially from the publication of this paper:

_____ Reimbursement for attending a symposium?

_____ A fee for speaking?

_____ A fee for organising education?

_____ Funds for research?

_____ Funds for a member of staff?

_____ Fees for consulting?

2. Have you in the past five years been employed by an organisation that may in any way gain or lose financially from the publication of this paper?

3. Do you hold any stocks or shares in an organisation that may in any way gain or lose financially from the publication of this paper?

4. Have you acted as an expert witness on the subject of your study, review, editorial, or letter?

5. Do you have any other competing financial interests? If so, please specify.

We are restricting ourselves to asking directly about competing financial interests, but you might want to disclose another sort of competing interest that would embarrass you if it became generally known after publication. The following list gives some examples.

    (a) A close relationship with, or a strong antipathy to, a person whose interests may be affected by publication of your paper.
    (b) An academic link or rivalry with somebody whose interests may be affected by publication of your paper.

    (c) Membership of a political party or special interest group whose interests may be affected by publication of your paper.

    (d) A deep personal or religious conviction that may have affected what you wrote and that readers should be aware of when reading your paper.

If you want to declare such a competing interest then please add it to your statement.

To learn more about the thinking that has led to this policy please read the editorial by Richard Smith BMJ 1998;317:291-2.

Please complete option 1 or 2 as appropriate and sign below

☐ 1. Please insert "None declared" under competing interests

or

☐ 2. Please insert the following statement under competing interests:

Title of paper:

Signature: _____

Print name too please: _____

Date_____

Example 6 (*Cont'd*)

(iv) The *International Journal of Epidemiology* (http://ije.oxfordjournals.org/) has a relatively simple 'Conflict of interest form' compared with the ones given above, and leaves it up to the authors to compose appropriate statements concerning any potential conflicts. It can be found at www.oxfordjournals.org/ije/for_authors/ije_conflict%20of%20interest%20form.pdf, and is reproduced as Example 7 with kind permission of Oxford University Press, which uses this basic form for several of its journals.

Conflict of Interest

OXFORD JOURNALS
OXFORD UNIVERSITY PRESS

COI_0506

International Journal of Epidemiology

Article title: _____

Author: _____   Ms number: dy _____

(Note: each author must complete a separate form and return it to the corresponding author)

Dear Author

IJE policy requires that all authors of all manuscripts sign a statement revealing (1) any financial interest in or arrangement with a company whose product was used in a study or is referred to in a manuscript; (2) any financial interest in or arrangement with a competing company, (3) any direct payment to an author(s) from any source for the purpose of writing the manuscript, and (4) any other financial connections, direct or indirect, or other situations that might raise the question of bias in the work reported or the conclusions, implications, or opinions stated—including pertinent commercial or other sources of funding for the individual author(s) or for the associated department(s) or organization(s), personal relationships, or direct academic competition. If the manuscript is published, such information may be communicated in a note following the text and references.

Please complete **Part A** *or* **Part B**.

A. Conflict of interest statement. (Sample statement)

I hold stock in *[business name]*, the makers of *[product]*, and am currently conducting research sponsored by this company. I am also a member of the speakers' bureau for *[business name]*.

My statement is as follows: _____

_____

_____

Printed name: _____

Signature: _____   Date: _____

B. I have had no involvements that might raise the question of bias in the work reported or in the conclusions, implications, or opinions stated.

Printed name: _____

Signature: _____   Date: _____

Example 7 Conflict-of-interest form for authors (as used by the *International Journal of Epidemiology* and reproduced with permission from Oxford University Press).

(v) The *American Journal of Gastroenterology* (www.amjgastro.com) asks its associate editors to disclose any relevant financial relationships and publishes the completed forms. The editors' completed 'Disclosure of relevant financial relationships' forms can be found at www.blackwellpublishing.com/pdf/AJC_combined.pdf and a blank form is reproduced as Example 8 with kind permission from the *American Journal of Gastroenterology*.

American Journal of Gastroenterology

ASSOCIATE & INTERNATIONAL ASSOCIATE EDITORS' DISCLOSURE OF FINANCIAL INTEREST STATEMENT

We, the editors of the *American Journal of Gastroenterology*, believe that all readers should be aware of any competing interests that the editors may have when considering an article or writing for the Journal.

Disclosure of Relevant Financial Relationships

List the names of proprietary entities producing health care goods or services, with the exemption of non-profit or government organizations and non-health care related companies, with which you or your spouse/partner have, or have had, a **relevant financial relationship*** within the past 12 months. Disclose only where the relationship is associated with the content of the activity.

☐ **No, I do not have a relevant financial relationship.**
☐ **Yes, I do have a relevant financial relationship**. Provide information below:

Nature of Relevant Financial Relationship (choose all that apply)	Name of Company(s)	Self	Spouse/ Partner
☐ Consultant		☐	☐
☐ Speaker's Bureau		☐	☐
☐ Grant/Research Support (Secondary Investigators need not disclose)		☐	☐
☐ Stock Shareholder (self-managed)		☐	☐
☐ Honoraria		☐	☐
☐ Full-time/part-time Employee		☐	☐
☐ Other (describe):		☐	☐

☐ By checking this box, I attest that the completed information is accurate.
Please accept this as my signature.

Printed Name _____
Date:

Example 8 Form for disclosure of associate editors' relevant financial relationships (reproduced with permission from the *American Journal of Gastroenterology*).

(c) Exclusive licence forms

Many journals and publishers no longer require authors to assign copyright for their articles to them – this remains with the authors. Instead, they ask that authors grant them an exclusive licence to publish their articles. In the forms used for this, details are given of the rights of authors to reuse their articles and the conditions that apply pre- and post-acceptance. The exclusive licence forms used by Blackwell Publishing are reproduced as Examples 9 and 10 by kind permission of Blackwell Publishing Ltd (www.blackwellpublishing.com). There are two versions, one for authors not opting to pay a fee to have their articles made available for free (open) access from day of publication (Example 9), and one for those authors choosing this option (Example 10).

(i) Non-open-access version

 Blackwell
Publishing

JOURNAL TITLE: Exclusive Licence Form

Author's name: ....

Author's address: ....

....

Title of article ("Article"): ....

....

Manuscript no. (if known): ....

Names of all authors in the order in which they appear in the Article: ....

....

In order for your Article to be distributed as widely as possible in <journal name> (the Journal) you grant Blackwell Publishing Ltd (Blackwell Publishing) an exclusive licence to publish the above Article <on behalf of society> including the abstract in printed and electronic form, in all languages, and to administer subsidiary rights agreements with third parties for the full period of copyright and all renewals, extensions, revisions and revivals. The Article is deemed to include all material submitted for publication with the exception of Letters, and includes the text, figures, tables, author contact details and all supplementary material accompanying the Article.

Please read this form carefully, sign at the bottom (if your employer owns copyright in your work, arrange for your employer to sign where marked), and return the ORIGINAL to the address below as quickly as possible. As author, *you remain the copyright owner of the Article*, unless copyright is owned by your employer. (US Federal Government authors please note: your Article is in the public domain.)

Your Article will not be published unless an Exclusive Licence Form has been signed and received by Blackwell Publishing.

Please note: You retain the following rights to re-use the Article, as long as you do not sell or reproduce the Article or any part of it for commercial purposes (i.e. for monetary gain on your own account or on that of a third party, or for indirect financial gain by a commercial entity). These rights apply without needing to seek permission from Blackwell Publishing.

- **Prior to acceptance:** We ask that as part of the publishing process you acknowledge that the Article has been submitted to the Journal. You will not prejudice acceptance if you use the unpublished Article, in form and content as submitted for publication in the Journal, in the following ways:
  - sharing print or electronic copies of the Article with colleagues;
  - posting an electronic version of the Article on your own personal website, on your employer's website/repository and on free public servers in your subject area.
- **After acceptance:** Provided that you give appropriate acknowledgement to the Journal, <Society> and Blackwell Publishing, and full bibliographic reference for the Article when it is published (see recommended statement below), you may use the accepted version of the Article as originally submitted for publication in the Journal, and updated to include any amendments made after peer review, in the following ways:
  - you may share print or electronic copies of the Article with colleagues;
  - you may use all or part of the Article and abstract, without revision or modification, in personal compilations or other publications of your own work;
  - you may use the Article within your employer's institution or company for educational or research purposes, including use in course packs;
  - [X months after publication] you may post an electronic version of the Article on your own personal website, on your employer's website/repository and on free public servers in your subject area.
  - Electronic versions of the accepted article must include the following statement, adapted as necessary for your Article:
  Author Posting. © The Authors {Insert year of publication} This is the author's version of the work. It is posted here for personal use, not for redistribution. The definitive version was published in {insert journal name}, {insert volume, issue number and pages}. http://dx.doi.org/ {insert doi number}

Please note that you are not permitted to post the Blackwell Publishing PDF version of the Article online.

All requests by third parties to re-use the Article in whole or in part will be handled by Blackwell Publishing. Any permission fees will be retained by the Journal. All requests to adapt substantial parts of the Article in another publication (including publication by Blackwell Publishing) will be subject to your approval (which is deemed to be given if we have not heard from you within 4 weeks of your approval being sought by us writing to you at your last notified address). Please address any queries to journalsrights@oxon.blackwellpublishing.com.

In signing this Agreement:

1. You hereby warrant that this Article is an original work, has not been published before and is not being considered for publication elsewhere in its final form either in printed or electronic form;

July 06

Example 9 Non-open-access exclusive licence form (reproduced with permission from Blackwell Publishing Ltd).

2.   You hereby warrant that you have obtained permission from the copyright holder to reproduce in the Article (in all media including print and electronic form) material not owned by you, and that you have acknowledged the source;

3.   You hereby warrant that this Article contains no violation of any existing copyright or other third party right or any material of an obscene, indecent, libellous or otherwise unlawful nature and that to the best of your knowledge this Article does not infringe the rights of others;

4.   You hereby warrant that in the case of a multi-authored Article you have obtained, in writing, authorization to enter into this Agreement on their behalf and that all co-authors have read and agreed the terms of this Agreement;

5.   You warrant that any formula or dosage given is accurate and will not if properly followed injure any person;

6.   You will indemnify and keep indemnified the Editors</Society> and Blackwell Publishing against all claims and expenses (including legal costs and expenses) arising from any breach of this warranty and the other warranties on your behalf in this Agreement.

By signing this Agreement you agree that Blackwell Publishing may arrange for the Article to be:

*   Published in the above Journal, and sold or distributed, on its own, or with other related material;
*   Published in multi-contributor book form or other edited compilations by Blackwell Publishing;
*   Reproduced and/or distributed (including the abstract) throughout the world in printed, electronic or any other medium whether now known or hereafter devised, in all languages, and to authorize third parties (including Reproduction Rights Organizations) to do the same;
*   You agree to Blackwell Publishing using any images from the Article on the cover of the Journal, and in any marketing material.

You authorize Blackwell Publishing to act on your behalf to defend the copyright in the Article if anyone should infringe it, although there is no obligation on Blackwell Publishing to act in this way.

As the Author, copyright in the Article remains in your name (or your employer's name if your employer owns copyright in your work).

Blackwell Publishing undertakes that every copy of the Article published by Blackwell Publishing will include the full bibliographic reference for your Article, together with the copyright statement.

Please tick only one of the boxes below.

☐ **BOX A: to be completed if copyright belongs to you**

☐ **BOX B: to be completed if copyright belongs to your employer (e.g. HMSO, CSIRO)**
The copyright holder grants Blackwell Publishing an exclusive licence to publish the Article including the abstract in printed and electronic form, in all languages, and to administer subsidiary rights agreements with third parties for the full period of copyright and all renewals, extensions, revisions and revivals.

Print Name of Copyright holder: ....
This will be printed on the copyright line on each page of the Article. It is your responsibility to provide the correct information of the copyright holder.

☐ **BOX C: to be completed if the Article is in the public domain (e.g. US Federal Government employees)** You certify that the Article is in the public domain. No licence to publish is therefore necessary.

Signature (on behalf of all co-authors (if any))   Print name: ....

....   Date: ....

If your employer claims copyright in your work, this form must also be signed below by a person authorized to sign for and on behalf of your employer, as confirmation that your employer accepts the terms of this licence.

Signature (on behalf of the employer of the author (s))   Print name:..

                                                             Print name of employer: ....

....   Date: ....

The rights conveyed in this licence will only apply upon acceptance of your Article for publication.

Data Protection: The Publisher may store your name and contact details in electronic format in order to correspond with you about the publication of your Article in the Journal. We would like to contact you from time to time with information about new Blackwell publications and services in your subject area. (For European contributors, this may involve transfer of your personal data outside the European Economic Area.) Please check the following boxes if you are happy to be contacted in this way:

☐ (conventional mailing)   ☐ (via e-mail)

Please return the signed form to: *Journal address*

July 06

Example 9 (*Cont'd*)

(ii) Open-access version

JOURNAL TITLE: Open Access Exclusive Licence Form

Author's name: ....

Author's address: ....

....

Title of article ("Article"): ....

....

Manuscript no. (if known): ....

Names of all authors in the order in which they appear in the Article: ....

....

In order for your Article to be distributed as widely as possible on an Open Access basis in <journal name> (the Journal) you grant Blackwell Publishing Ltd (Blackwell Publishing) an exclusive licence to publish the above Article <on behalf of society> including the abstract in printed and electronic form, in all languages. This will enable Blackwell Publishing to provide the benefits of Open Access publishing while preserving the integrity and archival status of all published articles.

The Article is deemed to include all material submitted for publication with the exception of Letters, and includes the text, figures, tables, author contact details and all supplementary material accompanying the Article.

Please read this form carefully, sign at the bottom (and if your employer owns copyright in your work, arrange for your employer to sign where marked), and return the ORIGINAL to the address below as quickly as possible. As author, *you remain the copyright owner of the Article*, unless copyright is owned by your employer. (US Federal Government authors please note: your Article is in the public domain.)

Your Article will not be published unless an Open Access Exclusive Licence Form has been signed and received by Blackwell Publishing, together with the accompanying Payment Authorization Form.

Open access: You require that the Article be published on an Open Access basis. This means that the full text of the Article can be accessed by readers without paying the publisher. <Society/Blackwell Publishing>'s current definition is at http://www.blackwellpublishing.com/static/openaccess.asp?site=1

Please note: You retain the following rights to re-use the Article, as long as you do not sell or reproduce the Article or any part of it for commercial purposes (i.e. for monetary gain on your own account or on that of a third party, or for indirect financial gain by a commercial entity). This does not affect your rights to receive a royalty or other payment for works of scholarship. These rights apply without needing to seek permission from Blackwell Publishing.

- **Prior to acceptance:** We ask that as part of the publishing process you acknowledge that the Article has been submitted to the Journal. You will not prejudice acceptance if you use the unpublished Article, in form and content as submitted for publication in the Journal, in the following ways:
    - o   sharing print or electronic copies of the Article with colleagues;
    - o   posting an electronic version of the Article on your own personal website, on your employer's website/repository and on free public servers in your subject area.
- **After acceptance:** Provided that you give appropriate acknowledgement to the Journal, <Society> and Blackwell Publishing, and full bibliographic reference for the Article when it is published, you may use the published version of the Article in the following ways:
    - o   you may share print or electronic copies of the Article with colleagues;
    - o   you may use all or part of the Article and abstract in personal compilations or other scholarly publications of your own work (and may receive a royalty or other payment for such work);
    - o   you may use the Article within your employer's institution or company for educational or research purposes, including use in course packs;
    - o   You may post the final PDF of the Article on your own personal website, on your employer's website/repository and on free public servers in your subject area. Blackwell Publishing will deposit the full-text of your article on publication with PubMed Central (PMC) in addition to any mirror of PMC (e.g. UKPMC). Electronic versions of the accepted Article must include a link to the published version of the Article together with the following text: 'For full bibliographic citation, please refer to the version available at www.blackwell-synergy.com'.

Third parties will be entitled to re-use the Article, in whole or in part, in accordance with the conditions outlined in the Creative Commons Deed, Attribution 2.5 Non-Commercial (further details from www.creativecommons.org), which allows Open Access dissemination of your work, but does not permit commercial exploitation or the creation of derivative works without your permission. Please address any queries to journalsrights@oxon.blackwellpublishing.com

20 December 2005

Example 10 Open-access exclusive licence form (reproduced with permission from Blackwell Publishing Ltd).

In signing this Agreement:

1. You hereby warrant that this Article is an original work, has not been published before and is not being considered for publication elsewhere in its final form either in printed or electronic form;
2. You hereby warrant that you have obtained permission from the copyright holder to reproduce in the Article (in all media including print and electronic form) material not owned by you, and that you have acknowledged the source;
3. You hereby warrant that this Article contains no violation of any existing copyright or other third party right or any material of an obscene, indecent, libellous or otherwise unlawful nature and that to the best of your knowledge this Article does not infringe the rights of others;
4. You hereby warrant that in the case of a multi-authored Article you have obtained, in writing, authorization to enter into this Agreement on their behalf and that all co-authors have read and agreed the terms of this Agreement;
5. You warrant that any formula or dosage given is accurate and will not if properly followed injure any person;
6. You will indemnify and keep indemnified the Editors</Society> and Blackwell Publishing against all claims and expenses (including legal costs and expenses) arising from any breach of this warranty and the other warranties on your behalf in this Agreement.

By signing this Agreement:
•   You agree that Blackwell Publishing may arrange for the Article to be published on an Open Access basis;
•   You agree to Blackwell Publishing using any images from the Article on the cover of the Journal, and in any marketing material;
•   You authorize Blackwell Publishing to act on your behalf to defend the copyright in the Article if anyone should infringe it, although there is no obligation on Blackwell Publishing to act in this way.

As the Author, copyright in the Article remains in your name (or your employer's name if your employer owns copyright in your work).

Blackwell Publishing undertakes that every copy of the Article published by Blackwell Publishing will include the full bibliographic reference for your Article, together with the copyright statement.

☐ **BOX A: to be completed if copyright belongs to you**

☐ **BOX B: to be completed if copyright belongs to your employer (e.g. HMSO, CSIRO)**
The copyright holder grants Blackwell Publishing an exclusive licence to publish the Article including the abstract in printed and electronic form, in all languages, and to administer subsidiary rights agreements with third parties for the full period of copyright and all renewals, extensions, revisions and revivals.

Print Name of Copyright holder: ....
This will be printed on the copyright line on each page of the Article. It is your responsibility to provide the correct information of the copyright holder.

☐ **BOX C: to be completed if the Article is in the public domain (e.g. US Federal Government employees)** You certify that the Article is in the public domain. No licence to publish is therefore necessary.

Signature (on behalf of all co-authors (if any))   Print name: ....

....   Date: ....

If your employer claims copyright in your work, this form must also be signed below by a person authorized to sign for and on behalf of your employer, as confirmation that your employer accepts the terms of this licence.

Signature (on behalf of the employer of the author (s))   Print name:................................. ....

        Print name of employer: ....

....   Date: ....

The rights conveyed in this licence will only apply upon acceptance of your Article for publication.

Data Protection: The Publisher may store your name and contact details in electronic format in order to correspond with you about the publication of your Article in the Journal. We would like to contact you from time to time with information about new Blackwell publications and services in your subject area. (For European contributors, this may involve transfer of your personal data outside the European Economic Area.) Please check the following boxes if you are happy to be contacted in this way:

☐ (conventional mailing)   ☐ (via e-mail)

Please return the signed form to: *Journal address*

20 December 2005

Example 10 (*Cont'd*)

Blackwell
Publishing

JOURNAL TITLE: Open Access Payment Authorization Form

In order for your article to be distributed as widely as possible on an Open Access basis in <journal name>, you must complete the accompanying Open Access Exclusive Licence Form in addition to this Payment Authorization Form. On payment of a one-off fee by or on behalf of the Author(s), your manuscript (as both full-text HTML and PDF) will be made available for viewing and download as an OnlineOpen article available through http://www.blackwell-synergy.com **permanently free of charge** to all readers – no subscriptions or memberships are required.

OnlineOpen papers receive the same rigorous editorial and production treatment as all other articles contained within the Journal, including the same quality of peer review, copy-editing and registration with Abstracting and Indexing services. OnlineOpen papers are also indicated as such in the print version of the journal, and with all third parties and indexing services supplied by Blackwell Publishing. Please note:

1.   You will only be charged if your paper is accepted for publication.
2.   Your paper will not be indicated as OnlineOpen to reviewers.

Blackwell Publishing cannot publish your paper as OnlineOpen without receipt of this form. The form must be completed together with the accompanying Open Access Exclusive Licence Form and sent to the Journal Editorial Office on or before the final acceptance of your manuscript.

Costs: | **There is a basic fixed fee of $2,500/€1,850/£1,250 for OnlineOpen**

UK Customers: Please add VAT at 17.5% (£1,468.75)

Other EU Customers:

- *If registered for VAT:* No VAT will be charged on completion of the details below
  VAT registration no:
  Name of organization:
- *If not registered for VAT:* Please add VAT at 17.5% (£1,468.75)

Customers outside the EU: No VAT will be charged

Your details:	
Title of article:	
Manuscript no (if known):	
Your name:	
Your address:	
Postcode/Zip:	
Email:	
Tel:	Corresponding author? Y[ ] / N[ ]
Fax:	*If no, please give name:*

Payment details Please debit credit card number:	☐☐☐☐☐☐☐☐☐☐☐☐☐☐☐☐
Expiry date (DD/MM/YY):	Verification ID Number: ☐☐☐☐ (3- or 4-digit number on reverse of Visa or front of Amex card)
Full name on Card:	
With the sum of: £/€/$ [delete as appropriate]	
Signature:	Date:
(Please note payment can only be accepted by credit card. We can accept Visa, MasterCard, Amex and Diners only)	
If you are asking a university or institution to pay for you, please give purchase order details below.	
Order No:	Date:

Please return your completed form to: *Journal address*

20 December 2005

Example 10 (*Cont'd*)

(d) Reviewing forms

Details are given below of reviewing forms from four journals and one society. They will give readers an idea of the types of forms that can be used and the sorts of questions that can be asked to help reviewers provide good and constructive reviews.

(i) The journal *The Holocene* (www.sagepub.co.uk/journalsProdDesc.nav?prodId= Journal201812) has a relatively simple form but contains questions to prompt responses on various aspects. The form is reproduced as Example 11 with kind permission from Sage Publications Ltd.

THE HOLOCENE: Peer review form

THIS FORM MAY BE SENT TO AUTHORS: Keep confidential statements separate

Paper title: ....
Author(s): ....

Checklist (place a cross in the box for any of the listed characteristics that need attention)

☐	Title:	Is it suitable? Does it reflect the content?
☐		Can it be improved?
☐	Abstract:	Is it informative? Are the main results and conclusions mentioned?
☐	Key words:	Are keywords provided?
☐	Introduction:	Are the aims clear?
☐	Methods:	Are they adequately explained?
☐	Results:	Are they described?
☐		Is any amplification or pruning necessary?
☐	Discussion:	Has this sufficient depth?
☐	Conclusions:	Do they follow from the evidence presented?
☐	Referencing:	Are there important missing references?
☐	Originality:	Does it contain sufficiently new results, ideas or techniques?
☐	Scope:	Is it interdisciplinary in scope and of more than local interest?
☐	Implications:	Is the broader context clear?
☐	Language:	Is it written in fluent English?
☐	Organization:	Is it well organized?
☐	Length:	Is it within the word limit (6000 for full Papers; 3000 for Reports)?
☐	Tables/Figures	Are there too many tables/figures (normal max. 8 for Papers; 4 for Reports)?

Figures: ☐ Are they necessary and well designed? (Where possible, figures should fit in a single column)
          ☐ Will they be legible on reduction?
          ☐ Is there a list of figure captions (on a separate sheet)?

Tables   ☐ Are they really necessary?
         ☐ Are they in the standard layout (see a recent issue of the journal)?

Overall evaluation (tick the most appropriate for your decision)

☐	*Excellent*	Presents an important new approach, new ideas or new information
☐	*Good*	Improves significantly on previous work of its type or contains new interesting information
☐	*Average*	Good work, but contains little novelty and may be of limited interest to most readers
☐	*Routine*	No errors, but likely to be of interest to local readers; not of international interest
☐	*Flawed*	Contains serious flaws in, e.g., project design, data analysis or presentation

Recommendation (tick whichever is applicable)

☐ Accept subject to minor revision, not requiring reconsideration by referees
☐ Possibly acceptable after moderate revision, possibly requiring reconsideration by referees
☐ Not accepted, major revision required, possibly with the opportunity of resubmission
☐ More suitable for publication in a different journal
☐ Not acceptable for publication in *The Holocene*

Print your name, if you wish it to be revealed to author(s) ....

On a separate sheet, please comment at length on any deficiencies of this paper, suggesting where improvements should be made.

Any confidential statements for the editor should be made in a separate letter

Example 11 Reviewing form used by *The Holocene* (reproduced with permission from Sage Publications Ltd).

(ii) The review scales reproduced as Example 12 with kind permission from Sage Publications Ltd are used by a number of journals at Sage Publications. Reviewers are asked to rate a number of criteria on a scale of 0 to 10, and are given guidance on what the high and low scores should denote.

Review of paper number:

This sheet is aimed to help you, the reviewer, to undertake a fair, complete and yet efficient review and is to help me, the editor, to decide on the fate of the paper. It is not intended to replace a free-text commentary which is always welcome, but it is intended to ensure some consistency. Any suggestions on it are most welcome.

Please rate the paper on each topic on the scale given.

Overall advice:

| 10 | 9 | 8 | 7 | 6 | 5 | 4 | 3 | 2 | 1 | 0 |

Accept (minor or no modification)   Reject (no merit whatsoever)

Readability and understandability

| 10 | 9 | 8 | 7 | 6 | 5 | 4 | 3 | 2 | 1 | 0 |

Good, minor editing only   Appalling, major rewrite needed

Abstract

| 10 | 9 | 8 | 7 | 6 | 5 | 4 | 3 | 2 | 1 | 0 |

A full and fair summary   Inaccurate &/or incomplete

Introduction

| 10 | 9 | 8 | 7 | 6 | 5 | 4 | 3 | 2 | 1 | 0 |

A fair review, justifies study   Biassed, incomplete, does not set the scene

Methods: description

| 10 | 9 | 8 | 7 | 6 | 5 | 4 | 3 | 2 | 1 | 0 |

Good description, easily understood and complete   Incomplete &/or badly described

Methods: measures used

| 10 | 9 | 8 | 7 | 6 | 5 | 4 | 3 | 2 | 1 | 0 |

Appropriate, good measures used   Inappropriate or no measures used

Methods: the design of the study

| 10 | 9 | 8 | 7 | 6 | 5 | 4 | 3 | 2 | 1 | 0 |

The design is good   The design is poor

Methods: analytical approach and statistical methods

| 10 | 9 | 8 | 7 | 6 | 5 | 4 | 3 | 2 | 1 | 0 |

Correct, and simplest possible for study   Incorrect or unecessarily complex

Results: presentation in text

| 10 | 9 | 8 | 7 | 6 | 5 | 4 | 3 | 2 | 1 | 0 |

Well presented and complete   Muddled or incomplete presentation

Results: presentation in tables & figures

| 10 | 9 | 8 | 7 | 6 | 5 | 4 | 3 | 2 | 1 | 0 |

Helpful, easily understood, complete   Muddled, incomplete, wasteful

Discussion: covers weaknesses of study

| 10 | 9 | 8 | 7 | 6 | 5 | 4 | 3 | 2 | 1 | 0 |

Fair discussion, placing in context   Fails to mention any weaknesses

Discussion: extrapolation and speculation

| 10 | 9 | 8 | 7 | 6 | 5 | 4 | 3 | 2 | 1 | 0 |

Reasonable conclusions, justified from data   Conclusions not justified at all

Example 12 Review scales (reproduced with permission from Sage Publications Ltd).

(iii) The *Hydrological Sciences Journal* (www.cig.ensmp.fr/~iahs/hsj/hsjindex.htm) has a comprehensive form that includes specific questions to prompt responses on a number of criteria. The form is reproduced as Example 13 with kind permission from the journal.

IAHS Press	web: www.iahs.info
Centre for Ecology and Hydrology	tel: +44 1491 692405
Wallingford	fax: +44 1491 692448/692424
Oxfordshire OX10 8BB, UK	e-mail: frances@iahs.demon.co.uk

Disseminating the results of hydrological research and practice worldwide

HYDROLOGICAL SCIENCES *JOURNAL* DES SCIENCES HYDROLOGIQUES

Editor: Professor Z. W. Kundzewicz

10 October 2006

Please return your review no later than *(six weeks deadline inserted here)*
Please contact Frances Watkins (frances@iahs.demon.co.uk) *if you are unable to provide a review, or meet the deadline.*

REFEREE'S REPORT: *Hydrological Sciences Journal* MS no.

Paper title:

Authors:

Referee name:

> *Please note that the contents of the manuscript remain confidential until published. Reviews are anonymous unless reviewers wish their names to be made known to the author(s). Would you like your name to be revealed to the author(s)?*

Aggregate assessment – How do you rate this paper in absolute terms?

Poor to fair	Good	Very good to excellent

Is the subject of the article	No	Possibly	Yes	Comments
Within the scope of the *Journal*?				

Please summarize, in one or two sentences, the main contribution and novelty, if any, of this paper:

Is the paper a new, original and valuable contribution to hydrological theory, methodology, modelling, education, etc?				
Is the paper a new, original, and valuable contribution to factual information about the hydrology of a particular region?				
If the reply to either of the above questions is positive, is the paper of sufficiently wide interest to merit publication in an international journal?				

Is the paper technically sound and free of errors of fact or logic?				
Are the objectives clear? Is the material clearly presented?				
Is the methodology appropriate?				
Are the assumptions and the analysis valid and adequately justified?				
Are the interpretations and conclusions sound and justified by the data?				
Are the data of appropriate quality?				

Example 13 Reviewing form used by the *Hydrological Sciences Journal* (reproduced with permission from the journal).

Referee's report HSJ   *(continued)*

	No	Possibly	Yes	Comments
Is the quality of the language satisfactory?				
Does the title of this paper clearly and sufficiently reflect its contents?				
Are the references adequate, up-to-date, and relevant?				
Are the approach, results and conclusions intelligible from the abstract alone?				
Are the key words informative, appropriate and complete?				
Are the illustrations of adequate quality, legible and understandable?				

Could the paper be shortened without detriment to the material presented in it (e.g. by removal of poor, irrelevant, excessive, or redundant material)? Please indicate such material in the manuscript. Are all illustrations and/or tables necessary? If not, could some of them be removed? Alternatively, could the information in the paper be more clearly or concisely conveyed by the use of tables or figures?

Please add any other specific comments you may have—if necessary, continuing on a separate sheet (sheets). Since the authors are requested to indicate on their revised papers where the reviewers' comments have been taken into account, it would help if you number any comments you may have.

Overall evaluation – The paper should be:

Accepted as it stands, apart from editorial changes.	
Accepted after minor revision.	
Subject to major revision. If revised paper is re-submitted, it needs to be reconsidered and re-reviewed.	
Rejected outright.	

The paper should be sent to another referee before terminating the review process (e.g. in the case of potentially contentious elements). If possible, please suggest the name (and e-mail) of a reviewer.	
If you have recommended major revision and re-submission, would you be willing to review the revised manuscript?	
Would you be willing to edit the language, should this paper be accepted?	

It is not necessary to return the paper itself unless you have inserted comments on it which have not been included in your reply. Referee reports sent as e-mail attachments are welcome—please save as "HSJXXXX review (your name)" in MS Word or Rich Text Format and send to frances@iahs.demon.co.uk.

Thank you in anticipation.

~ 10/10/2006

Example 13 (*Cont'd*)

(iv) The *International Journal of Psychoanalysis* (www.ijpa.org) has a special form that it sends to its reviewers when they are assessing *revised* manuscripts. The form is reproduced as Example 14 by kind permission of the editors of the *International Journal of Psychoanalysis.*

<u>**INTERNATIONAL JOURNAL OF PSYCHOANALYSIS**</u>

Revised Manuscript:

This is a REVISION of a manuscript that you have previously reviewed. Please answer each of these questions.

Please let us know if you are unable to review this manuscript.

A.   Has the author responded to your suggestions?   ___Not at all ___Some ___Almost all ___Totally

B.   Please tell us the current status of the following:

		Worsened	Remains the same	Improved
(i)	Title	_____	_____	_____
(ii)	Abstract	_____	_____	_____
(iii)	Organization	_____	_____	_____
(iv)	Clinical material	_____	_____	_____
(v)	Main argument(s)	_____	_____	_____
(vi)	Bibliography	_____	_____	_____

C.   Please indicate for which section(s) this paper is suitable.

Psychoanalytic Theory and Technique	Research
The History of Psychoanalysis	Educational and Professional Issues
Clinical Communications	Psychoanalytic Psychotherapy
Methodology	Interdisciplinary Studies

D . What is your current rating of the overall publishability?

Unacceptable 1..............2..................3..................4................5 Superior

Recommendation

___ Reject ___Needs a lot more work ___Needs a fair amount of work ___ Needs a little work ___Accept

Example 14 Reviewing form for *revised* manuscripts used by the *International Journal of Psychoanalysis* (reproduced with permission from the editors of the journal).

(v) Many journals do not provide general access to the reviewing forms they ask their reviewers to fill in for the manuscripts they review for them. The forms used by the journals of the Royal Society of Chemistry (www.rsc.org) are, however, freely available on its website and can be found at www.rsc.org/Publishing/ReSourCe/ rfreport/index.asp. The forms cover not just the different journals, but also the different article types considered.

(e) Reviewer questionnaire

The Royal Society of Chemistry (www.rsc.org) asks potential reviewers to fill in a 'Referees' questionnaire'. A very detailed checklist of subject categories broken down into specific keyword entries is provided to help reviewers submit a comprehensive summary of their research activities, expertise and interests. The full questionnaire can be found at www.rsc.org/Publishing/ReSourCe/RefereeGuidelines/Referees Questionnaire/index.asp. The first few pages are reproduced as Example 15 by kind permission of the Royal Society of Chemistry. The keyword entries go on for another six pages.

RSC | Advancing the
      | Chemical Sciences

The Royal Society of Chemistry,
Thomas Graham House,
Science Park,
Milton Road,
Cambridge,
UK CB4 0WF
Telephone: +44 (0)1223 420066
Fax: +44 (0)1223 420247

Referees' Questionnaire

Dear Colleague

Please complete the following questionnaire if you are willing to act as a referee for the Society's primary publications.

Please provide a brief summary on page 2 of your current or recent research activities and areas of expertise, and tick the keyword entries on pages 4–11 that cover your main interests. Both the textual and the keyword information are used in our search procedures, and the efficient operation of our system depends on both parts of the form being completed. If none of the categories cover your interests, please give a brief description on page 2.

Your help in identifying colleagues who would be willing and able to act as referees for The Society would also be appreciated.

Please return the completed questionnaire using the pre-paid reply label provided.

Thank you for your help and co-operation.

Robert J Parker
General Manager, Journals & Reviews

Name: (Prof/Dr/Mr/Mrs/Miss/Ms) ....

Address: ....

....

....   Postcode/Zip Code: ....

Telephone: ....   Fax: ....

E-mail: ....   Web address: ....

If you DO NOT wish to receive manuscripts for review electronically, please tick this box   ❏

The details provided in this questionnaire form will be input into our computerised administration system. There will be no disclosure of these details to anyone outside the RSC.

For non-members of the Royal Society of Chemistry:
The RSC will store the information you supply on its electronic records in order that information about its activities, products and services may be sent to you. If you DO NOT want to receive this information, please tick this box   ❏

Example 15 Questionnaire used by the Royal Society of Chemistry to obtain information from its reviewers on their research activities, expertise and interests (reproduced with permission from the Society).

Please summarise your current research activities (bullet points are clearest) in which you feel you would be able to act as a referee. Please use a separate sheet if necessary. Our refereeing system includes a facility for searching the text in this section and therefore the information provided here is an essential part of our searching procedures. It is particularly important that keywords covering your interests which do not appear elsewhere on the questionnaire are included here.

Other possible referees within your field of interest (see preamble; names and addresses please).

Any other information.

Example 15 (*Cont'd*)

This form is broken down into the following subject categories:

Example 15 *(Cont'd)*

Periodic Table Please encircle elements or groups of major/particular interest

1	2	3	4	5	6	7	8	9	10	11	12	13	14	15	16	17	18
(H)																	He
Li	Be											B	C	N	O	F	Ne
Na	Mg											Al	Si	P	S	Cl	Ar
K	Ca	Sc	Ti	V	Cr	Mn	Fe	Co	Ni	Cu	Zn	Ga	Ge	As	Se	Br	Kr
Rb	Sr	Y	Zr	Nb	Mo	Tc	Ru	Rh	Pd	Ag	Cd	In	Sn	Sb	Te	I	Xe
Cs	Ba	Lu	Hf	Ta	W	Re	Os	Ir	Pt	Au	Hg	Tl	Pb	Bi	Po	At	Rn
Fr	Ra	Lr	Rf	Db	Sg	Bh	Hs	Mt									
Lanthanides			La	Ce	Pr	Nd	Pm	Sm	Eu	Gd	Tb	Dy	Ho	Er	Tm	Yb	
Actinides			Ac	Th	Pa	U	Np	Pu	Am	Cm	Bk	Cf	Es	Fm	Md	No	

Chemistry
- ❑ Analytical science
- ❑ Biochemistry
- ❑ Bioinorganic chemistry
- ❑ Bioorganic chemistry
- ❑ Biophysical chemistry
- ❑ Biotechnology
- ❑ Chemical biology
- ❑ Chemical physics
- ❑ Crystallography
- ❑ Environmental chemistry
- ❑ Food chemistry
- ❑ Geology
- ❑ Geochemistry
- ❑ Green chemistry
- ❑ Industrial chemistry
- ❑ Inorganic chemistry
- ❑ Materials chemistry
- ❑ Mathematics
- ❑ Medicinal chemistry
- ❑ Miniaturisation
- ❑ Organic chemistry
- ❑ Organometallic chemistry
- ❑ Pharmaceutical science
- ❑ Physical chemistry
- ❑ Physical organic chemistry
- ❑ Physics
- ❑ Theoretical chemistry

Reactions
- ❑ Baeyer–Villiger
- ❑ Claisen
- ❑ Curtius
- ❑ Diels–Alder
- ❑ Fischer–Tropsch
- ❑ Friedel–Crafts
- ❑ Fries
- ❑ Heck
- ❑ Mannich
- ❑ Michael
- ❑ Mitsonobu
- ❑ Nazarov

- ❑ Peterson
- ❑ Sharpless
- ❑ Staudinger
- ❑ Vilsmeier
- ❑ Wittig

Reaction types
- ❑ Aldol
- ❑ Addition
- ❑ Cycloaddition
- ❑ Elimination
- ❑ Homolysis
- ❑ Oxidation
- ❑ Rearrangement
- ❑ Redox reactions
- ❑ Reduction
- ❑ Ring expansion/contraction
- ❑ Substitution

- ❑ **Organic chemistry**
- ❑ Alicyclic
- ❑ Aliphatic (general)
- ❑ N-containing aliphatic
- ❑ S-containing aliphatic
- ❑ Alkaloids
- ❑ Alkenes
- ❑ Alkanes
- ❑ Amino acids
- ❑ Aromatic
- ❑ Biomimetic
- ❑ Biosynthetic
- ❑ Carbohydrates
- ❑ Monosaccharides
- ❑ Polysaccharides
- ❑ Enzymes (general)
- ❑ Artificial enzymes
- ❑ Enzyme models
- ❑ Enzymes in synthesis
- ❑ Fullerenes
- ❑ Heterocyclic (general)
- ❑ Heterocyclic nitrogen
- ❑ Heterocyclic oxygen

- ❑ Heterocyclic sulfur
- ❑ Heterocyclic other element (please specify)
- ❑ Single ring heterocyclic
- ❑ Multi ring heterocyclic
- ❑ Single heteroatom type
- ❑ Multi heteroatom type
- ❑ Lipids
- ❑ Phospholipids
- ❑ Mechanistic studies
- ❑ Metabolites
- ❑ Animal metabolites
- ❑ Fungal metabolites
- ❑ Insect metabolites
- ❑ Marine metabolites
- ❑ Plant metabolites
- ❑ Natural products (general)
- ❑ Natural products synthesis
- ❑ Natural products structure elucidation
- ❑ Nucleic acids
- ❑ Nucleosides/nucleotides
- ❑ Peptides
- ❑ Polycyclic
- ❑ Polyketides
- ❑ Proteins
- ❑ Metalloproteins
- ❑ Quinones/phenolic compounds
- ❑ Reactive intermediates
- ❑ Carbenes
- ❑ Nitrenes
- ❑ Reactivity parameters (*e.g.* Hammett)
- ❑ Rings: 3/4-membered
- ❑ Rings: 7/8/9-membered
- ❑ Steroids
- ❑ Terpenes
- ❑ Vitamins
- ❑ Ylides

Organic: functional groups
- ❑ Acids and derivatives
- ❑ Alcohols/phenols
- ❑ Amines/imines and analogues

Example 15 (*Cont'd*)

- Carbonyl (aldehydes/ketones)
- Cationic (*e.g.* onium)
- Cyanides/isocyanides
- Diazo/azides
- Ethers/sulfides
- Halides and pseudohalides
- Hydroperoxides/peroxides
- Nitro/nitroso compounds
- Organoboron
- Organofluorine
- Organophosphorus
- Organosilicon
- Organo other element (please specify)

- **Organic synthesis**
- Asymmetric synthesis
- Computer-aided synthesis
- Microwave synthesis
- Organometallic reagents
- Regiocontrol
- Stereocontrol
- Supported reagents organic synthesis
- Synthetic methods and reagents

- **Pharmaceutical science**
- Antisense agents
- Cancer related chemistry
-   Anticancer drugs
-   Carcinogens
- Combinatorial chemistry
- DNA/RNA cleavage
- Drug delivery
- Drugs (preparation *etc.*)
-   Antibiotics
-   Antifungal drugs
-   Antitumour drugs
-   Antiviral drugs
- Gene delivery
- Pharmaceutical analysis
-   Antibiotics/penicillin
-   Drugs/formulations
-   Steroid drug
-   Residues
- Pharmacology
- Structure-activity relationships
- Structure-based drug design

- **Supramolecular chemistry**
- Calixarenes
- Catenanes
- Crystal engineering
- Cyclophanes
- Hydrogen bonding
-   Weak interactions
- Inclusion
-   Crown ethers
-   Cryptands
-   Cyclodextrins
- Donor–acceptor systems
- Macrocycles

- Macromolecules
- Molecular recognition
- Receptors
- Rotaxanes

- **Stereochemistry**
- Chirality
-   Resolution
- Optical activity
- Inorganic stereochemistry
- Organic stereochemistry
- Conformational analysis
- Theoretical stereochemistry
- Tautomerism

- **Inorganic chemistry**
- Cage compounds
- Coordination chemistry
- Main group chemistry
- Main group ring systems
- Transition metal chemistry
- Organometallic chemistry
- Inorganic synthesis
- Alkene ligands
- Alkyl ligands
- Alkyne ligands
- Allyl ligands
- Alums
- Arene ligands
- Aquation
- Bioinorganic
- Metals in biological systems
- Model bioinorganic
- Carbaboranes
- Carbene complexes
- Carbonyl complexes
- Chelates
- Clusters
- Crystal field theory
- Cyclometallated compounds
- Cyclopentadienyl ligands
- Defect structures
- Di/trinuclear compounds
- Fluxionality
- Inorganic solid state chemistry
- Halogens/halogen complexes
- Heterocyclic ligands
- Heteropolyanions
- Isopolyanions
- Macrocyclic ligands
- Magnetism (inorganic)
- Metal atom synthesis
- Metallocenes/sandwich compounds
- Metal–metal bonds
- Mixed valence compounds
- Nitrosyls
- Phosphine ligands
- Pi-bonding ligands
- Porphyrins and analogues
- Schiff's-base complexes
- Sigma-bonding ligands
- Silicates

- **Theoretical chemistry**
- *Ab initio* calculations
- Atomic & molecular properties
- Bond properties
-   Electron affinity
-   Ionisation energy
-   Molecular association
-   Potential energy surfaces
- Atomistic simulation
- Brownian motion
- Calculations on large systems
- Calculations on small systems
- Computational developments
- Computer graphics
- Computing in chemistry
- Density functional theory
- Dipole moments
- Electronic structure
-   Atomic orbitals
-   Band structure
-   Electron configuration
-   Electron correlation
-   Energy levels
-   Fermi levels
-   Frontier orbitals
-   Molecular orbitals
-   MO calculations
-   Wavefunctions
- Fractals
- Graph theory
- Intermolecular forces (theory)
- Intermolecular potentials
- Lasers (general)
- Liquid-structure theory
- MINDO/CNDO etc.
- Molecular dynamics
- Molecular mechanics
- Molecular modelling
- Monte Carlo calculations
- Polarisability
- Population analysis/charge density
- Tunnelling
- Quantum mechanics
- Quasiclassical dynamics
- Reaction dynamics
- Rydberg state
- Semiempirical calculations
- Spin coupling
- Spin properties
- Statistical thermodynamics
- Superfluidity
- Symmetry
-   Group theory
-   Point groups
-   Space groups
- Valence bond theory
- Trajectory calculations
- van der Waals complexes
- Virial/hypervirial theorems

- **Kinetics**
- Activation parameters

Example 15 (*Cont'd*)

(f) Suggested and non-preferred reviewer form

Some journals allow authors on submission to suggest individuals as potential reviewers for their manuscripts and also to name any that they would prefer were not approached to act as reviewers. A basic form for this is shown in Example 16.

Suggested reviewers

Authors are permitted to suggest up to six potential reviewers for their manuscripts. Suggested reviewers should not have been advisors, advisees or collaborators within the past 3 years.

Reviewer 1	**Reviewer 2**	**Reviewer 3**
First Name	First Name	First Name ....
Last Name	Last Name	Last Name ....
Institution	Institution	Institution ....
Telephone	Telephone	Telephone ....
Email	Email	Email ....
Reviewer 4	**Reviewer 5**	**Reviewer 6**
First Name	First Name	First Name ....
Last Name	Last Name	Last Name ....
Institution	Institution	Institution ....
Telephone	Telephone	Telephone ....
Email	Email	Email ....

Non-preferred reviewers

You have the option of suggesting up to two potential reviewers whom you would prefer were not chosen to review your manuscript. IF YOU TAKE UP THIS OPTION, YOU MUST STATE YOUR REASONS IN YOUR SUBMISSION LETTER. IF REASONS ARE NOT PROVIDED, YOUR REQUEST WILL NOT BE CONSIDERED.

Reviewer 1	**Reviewer 2**
First Name	First Name ....
Last Name	Last Name ....
Institution	Institution ....
Telephone	Telephone ....
Email	Email ....

Example 16 Form for authors to fill in on manuscript submission indicating any suggested and/or non-preferred reviewers.

(3) Guidance for reviewers

(a) The Royal Society of Chemistry (www.rsc.org) publishes a number of journals. It manages to provide comprehensive guidance to all its reviewers in a single document — 'Refereeing procedure and policy' — whilst giving information on the scope and specific requirements of the individual journals. The document can be found at www.rsc.org/Publishing/ReSourCe/RefereeGuidelines/RefereeingProcedureAndPolicy/index.asp and is reproduced as Example 17 by kind permission of the Royal Society of Chemistry.

Refereeing Procedure and Policy

Refereeing Procedure and Policy for Journals Published by the Royal Society of Chemistry†

Also see: www.rsc.org/authorguidelines

CONTENTS

† For more detailed information on this topic, as well as links to useful websites and software resources, see: http://www.rsc.org/resource.

1.0 Introduction

This document summarises the procedure used for assessing Primary Articles and Communications submitted to Journals published by the Royal Society of Chemistry (RSC).

2.0 The Journals

Submissions to the journals vary according to their subject matter, format, and type of article being presented. If it is felt that an article would be published more appropriately in an RSC journal other than the one suggested by the author, the referee should inform the Editor.

Some journals publish Communications (see Section 9.0, Communications), Letters, Comments and/or Opinions (see Section 7.0, Letters, Comments and Opinions).

2.1 The Analyst

The Analyst covers the theory and practice of all aspects of analytical and bioanalytical science. The journal only has space to publish three out of every ten submissions it receives. In order for a manuscript to be acceptable for *The Analyst*, it must report significant advances in analytical science.

2.2 Chemical Communications

A forum for preliminary accounts of original and significant work, in any area of chemistry that is likely to prove of wide general appeal or exceptional specialist interest (see also Section 9.0, Communications). Only a fraction of research work warrants publication in *Chemical Communications*, as the current rejection rate is 60%, and strict refereeing standards should be applied. Acceptance should only be recommended if the content of the paper is of such urgency or impact that rapid publication will be advantageous to the progress of chemical research.

Example 17 Guidance to reviewers provided by the Royal Society of Chemistry (reproduced with permission from the Society).

2.3 CrystEngComm
An electronic-only journal covering all areas of crystal engineering, including theoretical crystal engineering, techniques in crystal engineering, target crystals and properties (see also Section 10.0, Electronic-only Journals).

2.4 Dalton Transactions
Covering all aspects of the chemistry of inorganic and organometallic compounds, including biological inorganic and solid-state inorganic chemistry; the applications of physicochemical techniques to the study of their structures, properties and reactions, including kinetics and mechanism; new or improved experimental techniques and syntheses. For a research work to be accepted for publication it must report high quality new chemistry and make a significant contribution to its field.

2.5 Green Chemistry
Chemical aspects of clean technology, reduction of the environmental impact of chemicals (and fuels) whether from improved production methods, formulation and delivery systems, the use of sustainable/renewable resources and product substitution. Methodologies and tools for evaluating the environmental impact of the above, such as life cycle analysis, environmental risk analysis and legislative issues surrounding green chemistry. In no circumstances should papers just report the 'green angle' of previously published work or work submitted elsewhere, all submissions must be original.

2.6 Journal of Analytical Atomic Spectrometry
The journal covers the development of fundamental theory, practice and analytical application of spectrometric techniques to elemental research. It publishes only six out of every ten manuscripts it receives and referees should not hesitate to reject work which they feel is not of the required standard.

2.7 Journal of Environmental Monitoring
Physical, chemical and biological research relating to the measurement, impact and management of contaminants in all natural and anthropogenic environments. The journal places special emphasis on atmospheric science and human health issues and on the interface of analytical science with disciplines concerned with natural and human environments.

2.8 Journal of Materials Chemistry
The chemistry of materials, particularly those associated with advanced technology; modelling of materials; synthesis and structural characterisation; physicochemical aspects of fabrication; chemical, structural, electrical, magnetic and optical properties; applications; bio-related materials.

2.9 Lab on a Chip
Miniaturisation research and technology: its applications in chemistry, biology, physics, electronics, clinical chemistry, fabrication, engineering and materials science. Authors should clarify the advantages of carrying out the described processes/reactions at the micro- or nano-scale as opposed to the macro scale and must interpret and explain all their observations rather than just reporting them.

2.10 Molecular BioSystems
An interdisciplinary journal publishing novel and significant research that is emerging at the interface between chemistry and biology. The journal is intended as a forum for accounts of the research and development at the interface between chemistry and the -omic sciences and systems biology, in particular research concerned with cellular processes, metabolism, proteomics and genomics, systems biology, drug discovery, biomaterials, and all techniques relevant to these subject areas. All manuscripts should be written in a manner that is accessible to those working in the traditional fields of chemistry and biology as well as those working at the interface of the two subjects. In particular, abbreviations or acronyms should be clearly defined where they first appear in the text.

2.11 New Journal of Chemistry
A forum for the publication of original and significant work, in any area of chemistry that is likely to prove of wide general appeal or exceptional specialist interest. Only a fraction of research work warrants publication in the journal, which has a rejection rate is 60% for Articles and over 70% for Letters, and strict refereeing standards should be applied. Acceptance should only be recommended if the content of the paper will be advantageous to the progress of chemical research.

2.12 Organic & Biomolecular Chemistry
The journal brings together molecular design, synthesis, structure, function and reactivity in one journal. It publishes fundamental work on synthetic, physical and biomolecular organic chemistry as well as all organic aspects of: chemical biology, medicinal chemistry; natural product chemistry; supramolecular chemistry; macromolecular chemistry; theoretical chemistry; and catalysis.

2.13 Photochemical & Photobiological Sciences
Any aspect of the interaction of light with molecules, supramolecular systems and biological matter, for example, elemental photochemical and photophysical processes, the interaction of light with living systems, how light affects health, the use of light as a reagent in synthesis, the use of light as a diagnostic tool and for curative purposes, and areas in which light is a cost-effective catalyst or alternative source of energy.

2.14 Physical Chemistry Chemical Physics
All aspects of physical chemistry, chemical physics and biophysical chemistry including: catalysis; clusters; colloid and interface science; computational chemistry and molecular dynamics; electrochemistry; energy transfer; gas-phase reactions; kinetics and dynamics; laser-induced chemistry; materials science; photochemistry and photophysics; macromolecules and polymers; nanosciences; quantum chemistry and molecular structure; radiation chemistry; reactions in condensed phases; solid-state chemistry; spectroscopy of molecules; statistical mechanics; surface science; thermodynamics; zeolites.

2.15 Soft Matter
For high quality interdisciplinary research into soft materials and complex fluids, with a particular focus on the interface between chemistry and physics. Papers that describe applications and properties of soft matter set in context to the relevant science are also welcomed, but emphasis should be on the science rather than on the applications and properties themselves. The scope includes original research on important synthetic and characterisation techniques, and on simulation and modelling of soft matter. All manuscripts should be written in a manner that is accessible to those working in the traditional fields of chemistry and physics as well as those working at the interface of the two subjects. In particular, abbreviations or acronyms should be clearly defined where they first appear in the text.

3.0 Procedure
The referees' reports constitute recommendations to the appropriate Editor, who is empowered to take final action on manuscripts submitted. The Editor is responsible for all administrative and executive actions, and is empowered to accept or reject papers. It is the Editor's duty to see that, as far as possible, agreement is reached between authors and referees; although the referees may need to be consulted again concerning an author's reply to comments, further refereeing will be avoided as far as possible.

3.1 Adjudication of disagreements
If there is a notable discrepancy between the reports of the two referees, or if the difference between authors and referees cannot be resolved readily, a third referee may be appointed as adjudicator. In extreme cases, differences may be reported to the appropriate Editorial Board for resolution.

When a paper is recommended for rejection, the Editor will inform the authors. Authors have the right to appeal to the Editor if they regard a decision to reject as unfair. The Editor may refer to the Editorial Boards any papers which have been recommended for acceptance by the referees, but about which the Editor is doubtful.

3.2 Anonymity
The anonymity of referees is strictly preserved from the authors, and reports should be couched in terms which do not disclose the identity of the writer. A referee should never communicate directly with an author, unless and until such action has been sanctioned by the Society, through the Editor.

Example 17 *(Cont'd)*

3.3 Confidentiality
A referee should treat a paper received for assessment as confidential material. If a referee needs to consult colleagues to help with the review, the referee should inform them that the manuscript is confidential, and inform the Editor. Information acquired by a referee from such a paper is not available for disclosure or citation until the paper is published.

4.0 Policy
The primary criterion for acceptance of a contribution for publication is that it must report high-quality new chemical science and make a significant contribution to its field. Papers that do not contain new experimental results may be considered for publication only if they either reinterpret or summarise known facts or results in a manner presenting an advance in chemical knowledge. Papers in interdisciplinary areas are acceptable if the chemical content is considered satisfactory.

Papers reporting results regarded as routine or trivial are not acceptable in the absence of other, desirable attributes.

Although short papers are acceptable, the Society strongly discourages the fragmentation of a substantial body of work into a number of short publications; such fragmentation is likely to be grounds for rejection.

The length of an article should be commensurate with its scientific content; however, authors are allowed latitude (consistent with reasonable brevity) in the form in which their work is presented. Figures and flow-charts can often save space as well as clarify complicated arguments. Certain length restrictions apply to some Communications (see Section 9.0, Communications).

If a paper as a whole is judged suitable for the Journal, minor criticisms should not be unduly emphasised. It is the responsibility of the Editor to ensure the use of reasonably brief phraseology, and to assist the author to present his/her work in the most appropriate format. However, referees should not hesitate to recommend rejection of papers which appear incurably badly composed.

It should be clearly understood that referees' reports are made in confidence to the Editor, at whose discretion comments will be transmitted to the author. To assist the Editor, referees are requested to indicate which comments are designed only for consideration, as distinct from those which, in the referee's view, require specific action or an adequate answer before the paper is accepted.

Referees may ask for sight of supporting data not submitted for publication, or for sight of a previous paper which has been submitted but not yet published. Such requests must be made to the Editor, not directly to the author.

See also the RSC's 'Ethical Guidelines for Publication in Journals and Reviews'.†

4.1 Authentication of New Compounds
Referees are asked to assess, as a whole, the evidence in support of the homogeneity and structure of all new compounds. No hard and fast rules can be laid down to cover all types of compounds, but the Society's policy is that evidence for the unequivocal identification of new compounds should wherever possible include good elemental analytical data; for example, an accurate mass measurement of a molecular ion does not provide evidence of purity of a compound and must be accompanied by independent evidence of homogeneity (e.g. HPLC). Low resolution mass spectrometry must be treated with even more reserve in the absence of firm evidence to distinguish between alternative molecular formulae. Where elemental analytical data cannot be obtained, appropriate evidence which is convincing to an expert in the field may be acceptable.

Spectroscopic information necessary to the assignment of structure should normally be given. Just how complete this information should be must depend upon the circumstances; the structure of a compound obtained from an unusual reaction or isolated from a natural source needs much stronger supporting evidence than one derived by a standard reaction from a precursor of undisputed structure.

Referees are reminded of the need to be exacting in their standards but at the same time flexible in their admission of

evidence. It remains the Society's policy to accept work only of high quality and to permit no lowering of standards.

4.2 Electronic Supplementary Information (ESI)
Referees are encouraged to suggest that appropriate material is placed with the RSC's Electronic Supplementary Information (ESI) Service rather than the printed journal. Any supporting material for the ESI service supplied upon submission should be refereed to the same standard as the article.†

4.3 Use of Colour
The use of colour and/or half-tones is permitted in cases where genuine clarification results; referees may also be asked to advise on this [Electronic-only journals have different guidelines concerning the use of colour (see Section 10.0, Electronic-only journals)].

4.4 Titles and Summaries
Referees should comment on titles and summaries with the following points in mind.

Titles of papers are used out of context by several organizations for current awareness purposes. To enable such systems to serve chemical scientists adequately, titles must be written around a sufficient number of scientific words carefully chosen to cover the important aspects of the paper.

Summaries should preferably be self-contained, so that they can be understood without reference to the main text.

5.0 Speed of Refereeing
The RSC is anxious to maintain and to reduce further if possible the publication times now being achieved. In this connection, referees should submit their reports with the minimum of delay and within the specified time, or inform the Editor immediately if this is not feasible. If possible, referees should supply their reports in electronic format via the RSC's website.† In these cases, there is no need for referees to send a printed version of their report or to return the manuscript unless they are requested to do so by the Editor.

6.0 Suggestions of Alternative Referees
The Editor welcomes suggestions of alternative referees competent to deal with particular subject areas. Such suggestions are particularly helpful in cases where referees consider themselves ill-equipped (in terms of specialist knowledge) to deal with a specific paper, and in highly specialized or new areas of research where only a limited number of experts may be available. If, in such a case, the alternative and the original referee work in the same institution, the manuscript may be passed on directly after informing the Editor.

7.0 Letters, Comments and Opinions

7.1 Letters in *Dalton Transactions*
These are a medium for the expression of scientific opinions and views normally concerning material published in that journal; it is intended that contributions in this format should be published rapidly. The Letters section is for scientific discussion, and is not intended to compete with media for the publication of more general matters such as *Chemistry World*, or for revision/updating of authors' own work. Only rarely should a Letter exceed one printed column in length (about 1–2 pages of typescript). Where a Letter is polemical in nature, and if it is accepted, a Reply will be solicited from other parties implicated, for consideration for publication alongside the original Letter.

7.2 Letters in *New Journal of Chemistry*
These are concise articles that report results of immediate interest to the chemistry community: they may be complete publications, though a subsequent full paper may be justified, and should contain a brief experimental section.

7.3 Opinions in *New Journal of Chemistry*
Opinions should normally be limited to topics closely related to chemical science. This can include topics that are highly focused as well as those of broader interest to the chemical community. An

Example 17 (*Cont'd*)

Opinion is not intended to be a description of a consensual point-of-view on a given topic but could raise the need for a counter-opinion. It is a short, refereed, citable article on a topic related to chemical science that normally reports no new data but presents an opinion, hypothesis or conjecture on a topic judged by the referees and editor to be of interest to the readership.

The format is intended to allow more leeway for conjecture than the traditional formats. It should not be used to report a proposal that could be readily tested by currently available methods and published as a standard article. Opinions also could cover more general subjects related to educational, ethical, philosophical or sociological concerns of the chemical community. It should contain nothing that the referees judge offensive.

Ideally, an Opinion should not be longer than one printed page although no strict constraint on the length will be implemented. It will have a one-sentence abstract as well as a limited list of references. An Opinion may lead to the submission of a counter-Opinion, although noncontroversial issues could also be of interest to the chemical community.

7.4 Comments in *PCCP* and *JAAS*

Comments are a medium for the discussion and exchange of scientific opinions normally concerning material published in *PCCP* or *JAAS*. Submitted Comments will normally be forwarded to the authors of the work being discussed, and these authors will be given the opportunity to submit a Reply for publication together with the Comment. For publication of a Comment or Reply, they must be judged to be scientifically significant and of interest to either the *PCCP* or *JAAS* readership. Comments will not normally exceed a length of one printed journal page. Publication will take place only when all parties have had an opportunity to respond appropriately.

8.0 Polemical Papers

If the Editor considers a manuscript to be polemical in nature then the author of the paper being criticised will, wherever possible, be sent a copy of the manuscript.

9.0 Communications

9.1 Relationship between Full Papers and Preliminary Reports (*e.g.* Communications)

In cases where a preliminary report of the work described in a submitted paper has been published (for example in *Chemical Communications*), referees should alert the editor to any excessive and unnecessary repetition of material; this can arise in connection with Communications journals in which the restrictions on length and the reporting of experimental data are less severe than those of *Chemical Communications*. Furthermore, the acceptability of the full paper must be judged on the basis of the significance of the additional information provided, as well as on the criteria outlined in the foregoing sections.

9.2 Contributions to *Chemical Communications*

In most cases the preliminary reports published in *Chemical Communications* should be followed up by full papers in other journals, providing detailed accounts of the work. Referees are requested to comment on the RSC journal in which such full papers should be published. It is Society policy that only a fraction of research work warrants publication in *Chemical Communications*, and strict refereeing standards should be applied. The benefit to the reader from the rapid publication of a particular piece of work before it appears as a full paper must be balanced against the desirability of avoiding duplicate publication. The needs of the reader, not the author, must be considered, and priority in publication should not be allowed to determine acceptability. Acceptance should be recommended only if, in the opinion of the referee, the content of the paper is of such urgency or impact that rapid publication will be advantageous to the progress of chemical research.

Communications should be brief and should not exceed three pages in the printed form including Tables and illustrations. Communications should not include lengthy introductions and discussion, extensive data, and excessive experimental details and conjecture. Figures and tables will only be published if they are essential to understanding the paper. Authors must supply experimental evidence to support the conclusions drawn in the paper as Electronic Supplementary Information. The referees should comment on this supporting information in their reports with particular emphasis on whether the information does support the conclusions drawn in the paper and whether any additional information should be requested from the authors.

The refereeing procedure for Communications is the same as that for full papers, except that rapidity of reporting is crucial in order to maintain rapid publication.

9.3 Communications submitted to *OBC, Dalton Transactions, PCCP, Journal of Materials Chemistry, JAAS, The Analyst* or *Journal of Environmental Monitoring*

Criteria for acceptance of Communications submitted to *OBC, Dalton Transactions, PCCP, Journal of Materials Chemistry, JAAS, The Analyst* or *Journal of Environmental Monitoring* are similar to those for contributions to *Chemical Communications*, except that the work will be of more specialist interest. For *OBC* and *Dalton* Communications inclusion of key experimental data is expected.

10.0 Electronic-only Journals

For the Society's electronic-only journals there are no restrictions concerning page length, the use of colour or the number of tables and figures; however, the overall article length should be commensurate with the novel scientific content presented.

The refereeing procedure is the same as that for papers in printed journals, with the exception that referees will not be sent a printed copy of the article to be refereed. Instead, the article will only be made available for refereeing electronically.

11.0 X-Ray Crystallographic Work

All papers containing X-ray crystallographic work will be refereed for their chemical interest, and all crystallographic determinations will be assessed. If the Editor considers it advisable, the paper may be sent to a specialist crystallographer for comment. Assessors of crystallographic determinations will not normally be expected to check values of structural parameters for publication (*e.g.* bond lengths and angles against atomic coordinates; this will be done after publication by the appropriate crystallographic data centre), but should still pay attention to the quality of the experimental crystallographic work.

Papers will often contain the information in their titles that an X-ray structure determination has been carried out. However, this is not obligatory, especially if the X-ray determination forms only a minor part. Summaries should normally contain this information.

A structure referred to in a Communication will normally be fully refined. The Communication can then be considered to fulfil the archival function, and the structure determination may not require further detailed assessment when presented as part of a full paper. In the full paper, the author's purpose will then be served by a simple reference back to the original communication. However, if the crystallography is discussed again at any length in the full paper, the data should be re-presented to the referees in full, and re-published if considered necessary.

There may be other cases when an author wishes to publish a full paper in which the result of a crystal structure determination is discussed, but in which details or extensive discussion are considered unnecessary. The crystallographer may even be omitted as a co-author (for example when the determination is carried out by a commercial company). If the author is able to show that this procedure is appropriate, it will be allowed provided that it does not lead to unnecessary fragmentation. However, the author must provide, as supplementary information, sufficient data relating to the crystal structure determination to allow a crystallographer to make sure that the point made is correct. The brief published description of the determination should be supplemented by appropriate reference to 'unpublished work'.

Example 17 (*Cont'd*)

(b) The journal *Clinical Rehabilitation* (www.sagepub.co.uk/journalsProdDesc.nav?prodId=Journal201806) provides good general guidance to its reviewers, along with prompts to promote the return of good reviews. The 'Guidelines for reviewers' can be found under 'Guidance given to reviewers' at www.sagepub.co.uk/journalsProdAnnounce.nav?prodId=Journal201806 and are reproduced as Example 18 by kind permission of Sage Publications Ltd.

Clinical Rehabilitation: guidelines for reviewers (8th October 2006)

Clinical Rehabilitation aims to publish articles of relevance to the day-to-day practice of rehabilitation. It is a **peer-reviewed journal**, and thus depends crucially upon the quality of the reviews of submitted articles made by a large number of independent reviewers. Your work given freely(!) is crucial to this process. You have been approached to give an **unbiased opinion**. This document gives some guidance.

The **three aims** of peer review are:

- to **help select** articles for publication in the journal, selection being based on:
  - the scientific merit and validity of the article and its methodology;
  - the relevance of the article to the clinical practice of rehabilitation;
  - the interest of the topic to the clinical reader; and
  - the understandability of the article itself.

- to **improve** the articles wherever possible.
  - which data and analyses should be presented, including suggesting further analyses
  - structure and presentation of the article (detailed comments are not necessary)

- To check against **malfeasance** within the scientific and clinical community
  - writing; plagiarism, duplicate publication etc
  - data; fabrication or alteration
  - ethical and legal; not respecting participants (undue risk or inducement)

You have **three responsibilities:** to the author, to the reader, and to the journal.

The author
The author will have worked hard to carry out and write up the research. A referee should:
- give a reasonably quick reply (preferably within four weeks). If this is not possible please inform the editor as soon as possible. Authors are naturally impatient.
- give adequate, clear reasons for any comments, suggestions or recommendations. References are not necessary but may occasionally help.
- avoid personal bias, reading the paper for its own content.
- be constructive, not destructive, suggesting ways of overcoming any criticisms made, or of otherwise improving the paper
- read the paper **as if** blind to its origin if you (think) you know who wrote it

The reader of the journal
The readers of the journal will (or should) expect articles to have been scrutinised for major errors. Clinical Rehabilitation is read by a very wide range of professions from a wide range of cultures and countries, with varying levels of expertise. Readers depend upon informed experts reviewing the paper. Therefore the referee should check that:
- the work is original (if it claims to be);
- the background information given is correct, reasonably complete and covers most relevant issues without undue (hidden) bias;
- **the design of the study, and the logic of the arguments made are coherent;**
- the authors discuss any weaknesses openly and adequately;
- the results are credible and internally consistent;
- the conclusions can reasonably be drawn from the results, and are credible;
- the references are appropriate and accurate (as far as you know or can judge).

Example 18 Guidelines provided for reviewers by *Clinical Rehabilitation* (reproduced with permission from Sage Publications Ltd).

The journal

The journal publishes articles 'free', and so the authors must be encouraged to be as succinct as possible. Please comment if:

- you think that the article can be shortened,
- you think that tables or figures are unnecessary,
- you have any other suggestions to shorten or improve the article.

Secondly, the journal wants to retain its reputation. It should avoid publishing articles that

- are scientifically invalid,
- are duplicate publications,
- seem to include or condone illegal or unethical behaviour,
- are disrespectful of others in any way
- are misleading or simply without content.

An approach to reviewing a paper

Each reviewer develops their own approach to the task, and this editor does not wish to constrain his reviewers to any fixed format. Some suggestions are given here especially for those new to the job.

Your **comments are anonymous**, in that only the editor knows your identity. This allows you to be honest, but requires you to be polite and unbiased. Unless you request otherwise your comments will usually be sent to the authors with a covering letter. You and your co-reviewer will receive copies of my letter to the author and of each other's review (anonymously).

When reading the paper please consider two perspectives:

- as a **representative reader,** considering whether you would read it and understand it.
- as a **scientist,** considering the validity of statements, and of the conclusions.

Both the author and the editor appreciate **free-text comments** because they draw attention to matters that concern you. It is helpful to start your free-text comments with a short summary (1-4 sentences; 2-5 lines) of the main message of the paper. This means that the editor can get a quick overview of the content of the paper, and it can also reassure the author that you have understood the article.

After that you may choose the approach that you find best both for yourself and given the paper and its content. When making comments please draw attention to any major ambiguities or errors in writing, but you do not need to make detailed editorial comments on grammar, spelling etc.

In order to help you, a **series of specific questions** are given below. These may help you in thinking about the paper. They offer you a structure that probably applies to most papers, though certainly not to all papers. They also offer you the opportunity to score the paper on different aspects (and our web system offers an easy method for recording scores). However you are not obliged to use either these headings or the scoring system. It is for guidance only.

Scoring (optional)

If giving a score, note that:

- For each question the default is '99' which means that you think the question **does not apply,** or you **do not want to give a score** for that question.
- Otherwise please choose a number between 0 and 10, where 0 is the worst and 10 the best.

Example 18 (*Cont'd*)

Some questions to consider

1	What is your **overall advice?**
	I would appreciate your view on the value of the paper **assuming that your suggestions are carried out**. In other words, what is the **potential** outcome for the paper.
	0 = reject totally; the paper has no merit and might even be misleading
	10 = must accept; an outstanding study

2	How easy was the paper to **read and understand?**
	This refers to your opinion as a reader of the journal. Any difficulties you experienced were not your fault; they indicate a need for better writing or presentation. It does not specifically apply to the actual use of words if the author is clearly not a native speaker of English. However if particular sentences or paragraphs are unclear or ambiguous, please draw attention to them.
	0 = appalling presentation; the paper needs almost completely rewriting, structure poor
	10 = well constructed, logical flow of ideas, easy to read and understand

3	Is the **abstract** a reasonable summary?
	The abstract should give all the vital information, including some actual results (data, in studies reporting data). If unstructured, we will ask for a structured abstract.
	0 = inaccurate, misleading, seriously incomplete
	10 = full, clear, accurate

4	Is the **introduction** satisfactory?
	The introduction should explain why the study is needed, and set it in the context of existing work largely through references. Introductions should be succinct. They will usually end by posing the question(s) or hypotheses being explored.
	0 = incomplete, illogical, biased or otherwise unsatisfactory introduction
	10 = justifies study well, and refers adequately but not excessively to existing work

5	Are the **methods described** clearly?
	The methods section should in principle allow full replication of the study. Could you follow it? Was it complete? Did you understand what was done? Is there unnecessary discussion?
	0 = poorly structured, unclear, missing information
	10 = well structured, complete yet concise, no significant omissions, not excessive

6	Are the **methods used** appropriate?
	This refers to the design, measures and analytic procedures actually used. Was the design appropriate? Were the measures used or data collected appropriate? Have the data been analysed in the most efficient way possible, not too complex but utilising the data to their full extent. Comments on statistical analyses are helpful, but not essential (if you have specific concerns mention them; we can obtain specialist advice).
	0 = inadequate or faulty method used, not allowing any conclusion to be drawn
	10 = best possible methods used

7	Are the **results** presented clearly?
	This primarily refers to the structure of the results section (its ordering and flow), and the appropriate mix of text, tables and figures. It also refers to the data presented in the text.
	0 = incomplete data, poorly ordered, not well presented, obvious errors
	10 = well set out, logical ordering, easy to understand, results relate to questions posed

Example 18 (*Cont'd*)

8 | Are the **tables and figures** appropriate and accurate?
This refers to the data presented in tables or figures. Are they easily understood? Should more or less data be put in tables? Are the figures helpful?

   0 = tables and figures add nothing or are misleading or inaccurate
  10 = best use of tables and figures, no major changes needed

9 | Is the **discussion well structured**?
A discussion should have a reasonably logical flow. Could you follow it? Did it cover all main point concerning you? Were conclusions reasonable, and not too ambitious given the study?

   0 = poorly written discussion, difficult to follow and making unwarranted claims
  10 = well written discussion, balanced and drawing reasonable conclusions

10 | Does the discussion cover the **main limitations and weaknesses** of the study?
This is vital. The authors should be more aware than most of the flaws in their study. No study is perfect!

   0 = does not mention or acknowledge any weaknesses or limitations
  10 = covers all the main limitations, and gives reasonable weight to them

11 | Is the **extrapolation and speculation** reasonable, balanced and relevant to rehabilitation?
The readers are interested primarily in the clinical practice of rehabilitation. Any discussion should at least mention how the study might influence clinical practice.

   0 = unreasonable, illogical, unjustified or irrelevant speculation
  10 = clinically relevant and well justified conclusions

12 | Are the **clinical messages** appropriate?
The clinical messages are, in essence, the bottom line conclusions. Are those given justified? Have any been missed? Has the author drawn conclusions logically, and emphasised the most important ones?

   0 = messages not at all justified or consistent with study and data
  10 = fully appropriate

13 | Do you have any concerns on **malfeasance** (ethical, scientific, authorial)?
If you have any significant concerns, please let the editor know. You may wish to do this in a confidential letter. The editor will always keep you informed of his responses and actions, and will consult you before disclosing your concerns, and will not disclose your name (unless you request that).

I hope this guidance is helpful. Please send comments and helpful suggestions to the editor.
Derick Wade
Editor Clinical Rehabilitation and Professor of Neurological Rehabilitation
Oxford Centre for Enablement
Windmill Road
OXFORD OX3 7LD, UK
Tel: +44-(0)1865-737306
Fax: +44-(0)1865-737309
Email: clinical.rehabilitation@sagepub.co.uk

Anyone interested in the peer review process for Clinical Rehabilitation should read an editorial:
Clinical Rehabilitation 2004;**18**:117-124

Example 18 (*Cont'd*)

(c) The journal *Polar Research* (www.blackwellpublishing.com/journal.asp?ref= 0800-0395&site=1), the journal of the Norwegian Polar Institute, provides concise but helpful guidance to its reviewers in its 'Referee guidelines'. These are reproduced as Example 19 with kind permission from *Polar Research*.

POLAR RESEARCH • *The Journal of the Norwegian Polar Institute* • REFEREE GUIDELINES

The editor relies on referees to help determine whether the submission: 1) is based on good science; and 2) is optimally presented.

Comments to the author and editor should be kept separate. Your report itemizing your criticisms and summarizing your evaluation will be transmitted—together with the marked-up manuscript (if you return one) —to the author as they stand. In your comments intended for the author, please do not make any specific statement about whether the paper should be accepted or rejected. In your comments to the editor, offer your recommendation, summarizing the major weaknesses of the paper and stating whether you think they can be rectified. Consider the following aspects of the manuscript:

Scientific quality
- Importance of the research question or subject studied. Is the topic of broad or narrow interest?
- Originality of the work. E.g., is the paper sound and useful, but not novel?
- Appropriateness of the methodology and analysis. E.g., is the research fundamentally marred by a poor experimental design? Was the methodology appropriate but the subsequent analysis inadequate? Would additional tests improve the work?
- Quality of the data. E.g., can the quality be judged on the basis of how the data are described in the manuscript?
- Soundness of the interpretations and conclusions. Are these strongly supported by the data?
- Relevance of the discussion.
- Comprehensiveness of the references. Has relevant comparative information been overlooked?

Presentation
- Appropriateness of the title.
- Quality of the abstract. Does it include an adequate summary of methods, results and conclusions?
- Writing standard. Is the exposition easy or difficult to follow?
- Organization of the paper and its length. E.g., should sections be condensed, broken up or rearranged? Can the paper be shortened without reducing its impact?
- Quality and usefulness of the illustrations and tables. Are they all necessary? How can they be improved? Are additional figures or tables needed?

You are not required to correct deficiencies of style or mistakes in grammar. However, please note errors the copyeditor might miss, such as misspelled place names, incorrect scientific names of organisms and inappropriate scientific terminology. Also, please suggest ways to shorten overly long manuscripts by targeting needless repetition and superfluous graphics.

Contents of unpublished manuscripts should be treated as confidential. For more information about *Polar Research*, including the journal's Instructions for Authors, see our website at www.blackwellpublishing.com/journal.asp?ref=0800-0395&site=1

The identity of a referee is not disclosed to the authors unless this is the referee's wish. **If you do not wish to remain anonymous, please state this clearly in your letter to the editor.** Otherwise, your desire for anonymity will be assumed and your name and affiliation expunged from all material passed on to the authors.

Evaluations may be returned via e-mail to goldman@npolar.no. Or post your report and marked-up manuscript to:

Dr Helle V. Goldman
Chief Editor, *Polar Research*
Norwegian Polar Institute
Polar Environmental Centre
NO-9296 Tromsø, Norway
Fax +47 77 75 05 01
Tel. +47 77 75 05 00

Example 19 Guidelines provided for reviewers by *Polar Research* (reproduced with permission from the journal).

(4) Editorial letters

This section includes examples of letters that editorial offices need to use at various stages of the peer-review process. Only the body of the letters is given, with [...] denoting places where journal- and manuscript-specific information needs to be added. Letters should also include, at the minimum, manuscript reference number, title and authors, the name and contact details of the person being addressed, and the name, role and contact details of the person writing the letter. The journal name should always be given.

Letter 1: Acknowledgement of manuscript receipt to corresponding author

Dear

Thank you for submitting your manuscript to [journal name]. Please use the manuscript reference number given above in all future correspondence. As corresponding author, you will receive all future communications about this manuscript.

If there are any errors in the manuscript details above or in your contact details, please let the journal know via [email address].

During the review stage, all questions about the status of your manuscript should be directed to [name], via [email address].

(Note: see Chapter 3, Box 3.1 for examples of sentences that can be used to ask authors for missing or inadequate items.)

Letter 2: Acknowledgement of manuscript receipt to all the co-authors

Dear

The above manuscript, for which you are listed as a contributing author, has been received by [journal name]. Future communications regarding this manuscript will be sent to the corresponding author only.

If you need to contact us about your manuscript for any reason, please be sure to quote the manuscript reference number.

Letter 3: Assignment to handling editor

Dear

This new manuscript has been submitted to [journal name] and the authors have selected you as the handling editor.

The abstract and authors' submission letter are attached, and the usual instructions follow below.

Special note from EO: there are some language problems with this manuscript but standard is OK for review.

****

1. Please log on to [online site URL] and go to your Handling Editor Centre, where you will find this manuscript listed in your new manuscript queue. If you cannot act as handling editor, please let the editorial office know as soon as possible.
2. If you wish to have this manuscript reviewed, please send the names of 5–6 possible reviewers, and any other relevant information, to [name]. Please use the direct email hot-link button for your message.
3. If you want to reject this manuscript without review, please select 'Make Recommendation'. Complete the Recommendation Form that appears, follow the instructions on the screen, and then click on 'Send letter' when you have finished your message.
4. Please let us know as soon as possible if you would like us to ask other editors their opinions on whether or not this manuscript should be accepted for review.

****

Letter 4: Invitation to review

Dear

The above manuscript has been submitted to [journal name] and the handling editor, [editor's name], would very much like it to be reviewed by you. Would you be able to do this for us please? The manuscript details and abstract are attached to this message.

In our effort to make our reviewing process as quick and efficient as possible, we would ask you to return your report to us as quickly as possible, ideally within 14 days, but please let me know if you could review but would need longer than this.

If you agree to review this manuscript, you will be notified shortly by email about how to access the manuscript via the journal's online review site and be provided with all the information you will need to review the manuscript.

If you are not able to review this manuscript, we would welcome suggestions for alternative reviewers. Please do not, however, approach other potential reviewers direct, as the submission of this manuscript should be kept confidential.

I would be grateful if you could let me have your reply as soon as possible.

Our expert reviewers are crucial in helping maintain our high standards and I would like to thank you in advance for any help you can provide.

Letter 5: Invitation to review a resubmission

Dear

You recently kindly reviewed manuscript [manuscript number] for this journal. The manuscript was not accepted for publication but the authors were invited to submit a new manuscript after significant revision. We have now received this and as you reviewed the original submission we would very much value your opinion of the resubmission. Would you be able to do this for us please? The manuscript's details and abstract can be found attached to this message.

In our effort to make our reviewing process as quick and efficient as possible, we would ask you to return your report to us as quickly as possible, ideally within 14 days, but please let me know if you could review but would need longer than this.

If you agree to review this manuscript, you will be notified shortly by email about how to access the manuscript via the journal's online review site and be provided with all the information you will need to review the manuscript.

If you are not able to review this manuscript, we would welcome suggestions for alternative reviewers. Please do not, however, approach other potential reviewers direct, as the submission of this manuscript should be kept confidential.

I would be grateful if you could let me have your reply as soon as possible.

Our expert reviewers are crucial in helping maintain our high standards and I would like to thank you in advance for any help you can provide.

Letter 6: Assigning manuscript to reviewer

Dear

Thank you very much for agreeing to review this manuscript for [journal name]. To ensure that we can process manuscripts as quickly as possible, it would be very helpful if you could return your comments to us within 14 days. If by any chance you find you need more time than this, please could you let me know. Details of how to access the manuscript and submit your review online are given below. Please also don't hesitate to contact us if you need any assistance or further information.

We would ask you to treat the submission of this manuscript and the contents and information contained in it as confidential. If you wish to seek further advice from anyone outside your immediate research team could you please contact me before you do so.

Please would you not include any recommendation on publication in your report for the authors. If you would like to make any comments regarding this, please include them in the confidential report for the Editor.

Please be aware that, as a general principle, if you access any websites your IP address, and so you, can be identified. If there are any websites within this manuscript that you need to access to obtain information/materials to enable you to carry out an accurate and thorough review but that might compromise your anonymity, please let the editorial office know. We will endeavour to provide you with the information you require by another means. Alternatively, you can access websites anonymously using an anonymizer service, for example http://www.anonymization.net/. However, we cannot endorse or recommend any particular service.

Thank you again for your help with the review of this manuscript. I look forward to receiving your comments in the next 2 weeks.

****

TO ACCESS THE MANUSCRIPT AND SUBMIT YOUR REVIEW ONLINE
[Full details of how to do this]

****

Letter 7: Late agreement to review from reviewer

Dear

Thank you very much for your positive response to my request for you to review the above manuscript. It had already been sent to two reviewers when we received your reply, so we won't now need to send it to you. I would, however, like to keep your name on file as an 'On Hold' reviewer for this manuscript, in case we run into problems with its review at a later stage. If this is not acceptable to you, please let me know.

Letter 8: Checking reviewer has accessed manuscript

(Note: this is effectively the first 'reminder' about a review; see Chapter 4, Box 4.2, for an example of a series of messages of increasing urgency for reviewers who do not return reviews despite receiving reminders.)

Dear

Thank you very much for agreeing to review the above manuscript for [journal name]. We sent you details of how to gain access to the manuscript on [date]. This email is just to check that you received this information, that there are no problems, and that you have been able to view the manuscript and associated material on the journal's online submission and review site: [URL].

If you have any queries or need any assistance, please don't hesitate to contact me.

I look forward to receiving your review shortly.

Letter 9: Acceptance for publication

Dear

You will be pleased to know that your manuscript has been accepted for publication. [Authors need to be given information on the following sorts of things:

1. Whether any items are missing or of inadequate quality and need to be provided before the production process can start. These may be components of the manuscript itself or accompanying forms or policy-related assurances.
2. Details of the production process, including contact details for enquiries during that time, and whether production tracking is available for authors.
3. When authors can expect to receive proofs and how these will be sent or accessed (for example, as a PDF file or by downloading from a website after notification).
4. Approximate publication time and whether advanced online publication ahead of issue compilation is available.
5. If an open-access option is available, details of what authors who want to take this up should do.]

Letter 10: Provisional acceptance with revision

Dear

Thank you for submitting your manuscript to [journal name]. It has now been assessed by expert reviewers and their comments are attached. They can also be viewed, along with the editorial correspondence, in your Author Centre on our online site [URL].

You will be pleased to know that your manuscript has been provisionally accepted for publication pending satisfactory revision.

[Details of what revisions authors need to do, which they do not need to do, and any other conditions that need to be taken into account.]

Please note that revised manuscripts must be received within 2 months of authors being notified of conditional acceptance. Full instructions for submitting your revised manuscript are given below.

Thank you for submitting this work to [journal name]. I'm glad we're able to bring you good news and I look forward to receiving the revised manuscript.

****

INSTRUCTIONS FOR SUBMISSION OF YOUR REVISED MANUSCRIPT
[Authors need to be given information on the following sorts of things:
1. Time limit if not given in actual letter.
2. Where and how to submit the revised manuscript.
3. Items that need to be included – see Chapter 5, Box 5.1, for a list.]

****

Letter 11: Rejection without external review

(a) Language problems – encourage resubmission

Dear

Thank you for submitting your manuscript to [journal name]. All new manuscripts are given a preliminary review by the Editors to assess whether the subject matter and general content are appropriate for this journal.

You will be pleased to know that your manuscript has passed this initial screen. However, we feel that the standard of [language] in your manuscript is below that needed for this journal and it will need to be improved before we can send your manuscript out for full review. As it stands, reviewers would find it difficult to make an accurate assessment of the science and this would not be fair to your work, or indeed to our reviewers. We are therefore returning the manuscript to you in your best interests, to give it the best chance of a full and fair review.

Thank you for considering [journal name] for publication. We hope you find our action helpful and we look forward to sending your manuscript out for full review if you do decide to submit an improved version in the near future.

(b) Language problems – no encouragement to resubmit

Dear

Thank you for submitting your manuscript to [journal name]. All new manuscripts are given a preliminary review by the Editors to assess whether the subject matter and general content are appropriate for this journal.

Unfortunately, the Editors felt that the study you report would not really be of enough general interest to our broad readership and it is better suited for a more specialized journal, for example [journal name] or [journal name]. In the spirit of helpfulness, we would advise that, before you submit to another journal, you go over the whole manuscript very carefully and improve the language or that you seek help with this. As the manuscript stands, reviewers would find it difficult to make an accurate assessment of the science, and this would not be fair to your work.

Thank you for considering this journal for publication. We are sorry to disappoint you on this occasion and wish you success in getting your work published in another journal.

(c) Inappropriate topic

Dear

Thank you for submitting your manuscript to [journal name]. All new manuscripts are given a preliminary review by the Editors to assess whether the subject matter and general content are appropriate for this journal.

Unfortunately, the Editors felt that the topic covered in your manuscript is outside the scope of this journal and is better suited for one that publishes papers in this area, for example [journal name] or [journal name].

Thank you for considering this journal for publication. We are sorry to disappoint you on this occasion and wish you success in getting your work published in a more suitable journal.

Letter 12: Rejection after external review with resubmission encouraged

(Note: appropriate level of enthusiasm for a resubmission needs to be conveyed at *.)

Dear

Thank you for submitting your manuscript to [journal name]. It has now been assessed by expert reviewers and their comments are attached. They can also be viewed, along with the editorial correspondence, in your Author Centre on our online site [URL].

Unfortunately, on the basis of the reviewers' comments and my own reading of your manuscript, we are not at this stage able to accept it for publication. However, if you are able to address the reviewers' comments and make the necessary revisions and improvements we would be happy to receive a new submission from you in the future. This would need to [full details of all the conditions that need to be fulfilled for a resubmitted manuscript to be considered, including any time constraints].

I'm sorry we're not able to bring you a positive outcome at this stage, but I hope our reviewers' comments will be helpful to you*

*   and that you will resubmit the manuscript to us in the near future.
*   in deciding the way forward for your manuscript.
*   in preparing your manuscript for submission to another journal if you are not able to make the revisions required for the manuscript to be reconsidered.

Letter 13: Rejection after external review with no resubmission encouraged

Dear

Thank you for submitting your manuscript to [journal name]. It has now been assessed by expert reviewers and their comments are attached. They can also be viewed, along with the editorial correspondence, in your Author Centre on our online site [URL].

I'm afraid that, based on the reviewers' comments and my own reading of your manuscript, we are not able to accept it for publication. Unfortunately, it seems very unlikely that any revisions could be made that would render the manuscript acceptable for publication in this journal because [reasons why]. We therefore recommend that you submit your manuscript to another journal and we hope that our comments are helpful to you in this.

Thank you for considering this journal for publication. We are sorry to disappoint you on this occasion and wish you success in getting your work published in another journal.

Letter 14: Thanks and notification of outcome to reviewers

Dear

Recently you kindly reviewed the above manuscript for [journal name]. I thought you would be interested to know the outcome of the review process and to see the reviewers' reports. The decision was [decision] and the reviewers' reports are attached.

Thank you very much for reviewing for us — we greatly appreciate your help and input.

Appendix III
Useful websites

This appendix lists various websites that will be of interest and use to people involved in journal editorial work. Many of the organizations have been mentioned in the main text, some a number of times. Here, a brief description is given of them and what they do. Most of the websites contain information about, and links to, other sites that readers will find helpful.

Association of Learned and Professional Society Publishers (ALPSP; www.alpsp.org)

ALPSP is the international trade association for not-for-profit publishers and those who work with them; it is also the largest association of scholarly and professional publishers in the world. It provides representation of the sector, professional development activities and a wealth of information and advice. It runs many very useful and informative seminars and offers a wide variety of training courses at different levels. The training courses are tailored to meet the requirements of the academic and professional publishing market and cover business and management, editorial, production and e-publishing, marketing, and legal issues. ALPSP regularly commissions and publishes reports on topics of direct relevance to the publishing industry. It also has a journal, *Learned Publishing*, which it publishes in collaboration with the Society for Scholarly Publishing (see this appendix, page 273), a monthly email newsletter, *ALPSP Alert*, and several very active and high-quality listserv discussion fora. There are local Chapters in various locations worldwide.

Committee on Publication Ethics (COPE; www.publicationethics.org.uk)

COPE is a voluntary organization formed in 1997, originally as a self-help group set up by biomedical editors for editors, to address breaches of research and publication ethics. It is 'a forum for editors of peer-reviewed journals to discuss issues related to the integrity of the scientific record; it supports and encourages editors to report, catalogue and instigate investigations into ethical problems in the publication process' (accessed 8 August 2006). One of its aims is to develop good practice. It provides guidelines on good publication practice and has put together a code of conduct for editors. COPE also advises on anonymous cases of misconduct; summaries of the case studies, with all the discussion points, can be found on its website.

CONSORT (www.consort-statement.org)

CONSORT stands for Consolidated Standards of Reporting Trials, but is always referred to by just the acronym. The CONSORT statement is a tool to improve the quality of reporting of randomized controlled trials (RCTs). It allows RCTs to be reported in a standard, transparent and evidence-based way. It comprises a checklist and a flowchart, which together are called CONSORT (e.g. '...use of CONSORT to...'). The checklist contains all the things that should be addressed in a trial report; the flow diagram shows the progress of all participants in the trial, from time of randomization until the end of their involvement. The aim is to clarify the experimental process and aid evaluation of the data. The CONSORT statement is an evolving document, so readers should refer to the website for the latest version. The checklist and flow diagram current in August 2006 are reproduced in Appendix II (see page 217).

Council of Science Editors
(CSE; www.councilscienceeditors.org)

The CSE used to be the Council of Biology Editors (CBE) but adopted the new name in 2000 to reflect more accurately its membership. Its mission is 'to promote excellence in the communication of scientific information' and its purpose is 'to serve members in the scientific, scientific publishing, and information science communities by fostering networking, education, discussion, and exchange and to be an authoritative resource on current and emerging issues in the communication of scientific information' (accessed 8 August 2006). There is much very helpful information in the CSE's extensive editorial policy statements. There are also very useful reference links to all sorts of resources, such as a large selection of dictionaries, grammar and style guides, maps, quotations, various databases, and lots of general information.

European Association of Science Editors
(EASE; www.ease.org.uk)

EASE is 'an internationally oriented community of individuals from diverse backgrounds, linguistic traditions and professional experience who share an interest in science communication and editing' (accessed 8 August 2006). It has an electronic forum for the exchange of ideas, and holds a major conference every 3 years. Its *Science Editors' Handbook* contains much useful information, divided up into sections on: (1) Editing, (2) Standards and Style, (3) Nomenclature and Terminology, and (4) Publishing and Printing. It is an evolving publication. All the chapters are held in a loose-leaf ring binder. They are periodically updated and new chapters are being added.

International Association of Scientific, Technical and Medical Publishers (STM; www.stm-assoc.org)

The mission of STM is 'to create a platform for exchanging ideas and information and to represent the interest of the STM publishing community in the fields of copyright, technology developments, and end user/library relations' (accessed 8 August 2006). STM includes large and small publishing companies, secondary publishers, and learned societies. Its membership accounts for an estimated 80% of the annual output of scientific research communication. There is access to a number of documents, statements and public correspondence on its website.

International Committee of Medical Journal Editors (ICMJE; www.icmje.org)

The ICMJE is made up of a group of editors from general medical journals who meet annually to discuss the 'Uniform Requirements for Manuscripts Submitted to Biomedical Journals: Writing and Editing for Biomedical Publication'. Guidelines for the format of manuscripts for general medical journals were first produced in 1979 following an informal meeting of a small group of editors of general medical journals in Vancouver, British Columbia, in 1978. The group became known as the Vancouver Group, and the guidelines are still sometimes referred to as the Vancouver Guidelines. The Vancouver Group evolved into the ICMJE, which has since produced many updates of the Uniform Requirements. These guidelines have been extended to cover more than just manuscript preparation, and now include ethical considerations and many editorial issues. All editors and authors will benefit from looking at them and will find them a very valuable resource. The ICMJE allows the Uniform Requirements to be reproduced for educational, not-for-profit purposes, and encourages their distribution.

International Council for Science (ICSU – acronym derived from the previous, pre-1998, name – the International Council of Science Unions; www.icsu.org)

The ICSU is a non-governmental organization with a global membership that includes both national scientific bodies and international scientific unions. The ICSU provides a forum for the discussion of issues relevant to international science policy and it actively advocates freedom in science, promotes equitable access to scientific data and information, and facilitates science education. It addresses global issues in partnership with other organizations and acts as an advisor on a wide range of topics from ethics to the environment.

International Publishers' Association (IPA; www.ipa-uie.org)

The IPA is a long-standing (established in 1896) non-governmental organization that represents the publishing industry, with consultative relations with the United Nations. Its membership is made up of national publishers' organizations and those organizations must be representative of their countries (there are no individual publishing company members). Each country is generally represented by only one organization.

The IPA's mission is:

- 'to uphold and defend the right of publishers to publish and distribute the works of the mind in complete freedom
- to promote and protect by all lawful means the principles of copyright
- to overcome illiteracy, the lack of books and of other education materials
- to assure the unrestricted import and export of books and other materials produced by publishers' (accessed 8 August 2006).

International Standard Randomised Controlled Trial Number Register (ISRCTN; http://isrctn.org)

The ISRCTN is a simple numeric system (based on randomly generated 8-digit numbers prefixed by ISRCTN) for the unique identification of randomized controlled trials worldwide. This enables all publications and reports resulting from a specific trial to be tracked. A pilot was introduced in March 2000, and the scheme was formally launched in May 2003. The website gives answers to frequently asked questions and readers are referred there for further details and up-to-date information. As mentioned in Chapter 3 (page 30), clinical trial registration is in the process of change and attempts are being made to introduce greater regulation and transparency into clinical trials.

Office of Research Integrity (ORI; http://ori.dhhs.gov)

The ORI is part of the Office of Public Health and Science within the Office of the Secretary of Health and Human Services in the US Department of Health and Human Services. It promotes integrity in biomedical and behavioural research supported by the US Public Health Service at around 4000 institutions worldwide. ORI monitors institutional investigations of research misconduct and promotes responsible conduct of research through educational, preventative and regulatory activities. It develops policies, procedures and regulations related to the detection, investigation and prevention of research misconduct. It also reviews and monitors investigations into research misconduct carried out by institutions whose research is funded by the Public Health Service and provides assistance when required. Helpful guidance documents can be found on its website.

Society for Scholarly Publishing (SSP; www.sspnet.org)

The mission of SSP is 'to advance scholarly publishing and communication and the professional development of its members through education, collaboration and networking among individuals in this field' (accessed 8 August 2006). It provides the opportunity for interaction among members in all aspects of scholarly publishing, including journal and book publishers, librarians, manufacturers, and web editors. It has links to many organizations and resources on its website, and includes email and telephone contact details for some of the listings. The scope is very broad-ranging but grouped under topic headings.

World Association of Medical Editors (WAME – pronounced 'whammy'; www.wame.org)

WAME is 'a voluntary association of editors from many countries who seek to foster international cooperation among editors of peer-reviewed medical journals' (accessed 8 August 2006). Membership is free and open to all editors of peer-reviewed medical journals. WAME aims particularly to assist editors in developing countries and editors of small journals, who may have problems attracting high-quality manuscripts, have limited funds and lack training and expertise in editing and publishing. WAME's website provides many resources that will be useful to all editors, not just those from medical journals – policy statements, ethical considerations, and guidance for editors before and after taking up editorial positions. It also has a comprehensive listing of its listserv discussions, with links to the postings.

Appendix IV
Alternative models of peer review

The peer-review process described in this book is usually considered the 'traditional' model and is still the one used by the great majority of scientific journals. Variations of this model and new systems have been suggested, and a number have been tried. This appendix gives a brief description of some of these and provides details of where interested readers can go to find out more. Readers are also directed to the 22 articles published online in 2006 by the journal *Nature* on various aspects of editorial peer review to accompany its peer-review trial (see this appendix, page 279) – links to all the articles can be found at www.nature.com/nature/peerreview/debate/index.html.

(1) Pre-submission peer review

A form of pre-submission peer review occurs naturally in most scholarly work. Authors informally seek feedback from colleagues and fellow scholars on their work and their draft manuscripts. In some areas this has become more formalized, and special 'pre-print' repositories or servers have been set up. The best known and most established is arXiv (pronounced 'archive' in English – the X standing for the Greek 'chi'; http://arxiv.org), which is described on its website as 'a highly-automated electronic archive and distribution server for research papers'. It was set up in 1991 by the physicist Paul Ginsparg at the Los Alamos National Laboratory in New Mexico as a free repository to enable physicists to exchange information (it was, until the end of 1998, known as xxx.lanl.gov). It is now based at Cornell University in Ithaca, New York, and houses (in 2006) close to 400,000 electronic research papers ('e-prints') in physics, mathematics, nonlinear science, computer science, and quantitative biology. Non-reviewed pre-prints of papers in these subjects can be posted on arXiv before being submitted for publication. Other researchers then comment on them and authors are able to make revisions to their papers before submitting them to regular peer-reviewed journals if they want, or they can just update their submissions. Most authors do go on to submit to journals and are supposed to post on arXiv the versions accepted for publication; previous versions remain available. The system works well in this area of science and has become accepted as the norm. However, it may not be suitable for all disciplines. For example, in fast-moving fields (such as molecular biology) authors may be inhibited from adopting such a system because of fear of other researchers quickly repeating their work, submitting to another journal, and so 'scooping' them.

(2) Post-publication review or commentary

Post-publication review or commentary is an area that a number of journals are experimenting with or adopting. The advent of online journals and sophisticated and flexible access platforms has made this both feasible and relatively easy to introduce. It isn't a substitute for peer review, but rather a supplement to it, and occurs after papers have been peer reviewed, accepted and published (readers are directed to an article by Stevan Harnad in which he discusses the concept of open peer commentary: 'The invisible hand of peer review', *Exploit Interactive*, 2000, issue 5, April, www.exploit-lib.org/issue5/peer-review/). Comments and responses submitted by readers about papers are posted online with those papers. Considerable thought needs to be put into how to set up a post-publication commentary system to prevent abuse or the posting of inappropriate comments. Most sites will, therefore, require some form of moderation and this may be beyond the resources of some journals. There are also concerns that comments might not be broadly representative – for example, junior researchers may be reluctant to openly post negative comments about senior researchers, and people who might have valuable comments to make may not be interested enough to do this or be prepared to devote the time required.

One publisher that invites post-publication posting of comments for its journals is BioMed Central (BMC; www.biomedcentral.com). BMC is an independent online publisher that provides open access to articles by levying article-processing charges that are paid by the authors, their institutions or their funding bodies. Authors can apply for waivers of the charge in cases of hardship. Readers are able to post comments, ask the authors questions, or just post additional information, and these postings are linked to the papers concerned and freely available. There is a 'Comments Policy' (www.biomedcentral.com/info/about/commentpolicy) and readers who are thinking about introducing post-publication commentary might want to take a look at this to get an idea of the sorts of things that need to be considered. For example, BMC requires individuals who want to post a comment to register as users, and comments are subject to editorial approval and moderation to prevent posting of anything that is indecent, offensive, potentially libellous, too trivial, incomprehensible, or advertising in disguise. All people posting comments must also complete a competing-interest declaration.

The *BMJ* (*British Medical Journal*; http://bmj.bmjjournals.com) also has post-publication commentary and allows readers to post 'Rapid Responses' to its content and these appear at the ends of the relevant contributions. The comments are moderated and contributors have to declare any competing interests. The journal enables readers to keep up to date and provides them with easy access to new responses by mounting all those received in the past day; those received within the past 2–21 days can also be accessed.

A slightly different form of post-publication review/commentary is operated by Faculty of 1000 Biology (F1000 Biology; www.f1000biology.com), where assessments and comments are posted not by the general readership but by a defined body of specialists. F1000 Biology is a subscription-based 'literature-awareness tool'. It systematically highlights and reviews what are considered to be the most interesting papers published in the biological sciences based on the recommendations of a group ('Faculty') of over 1000 selected researchers (hence the name 'Faculty of 1000'). These researchers are divided up into subject 'faculties', which in turn are subdivided into 'sections' which cover all of biology, and they select papers based on merit and not because of the journals in which they are published. Each member is asked to evaluate and provide structured comments on the 2–4 most interesting papers they read each month. They assign each paper a rating: 'recommended' (rating of 3), 'must read' (rating of 6), or 'exceptional' (rating of 9). Papers may be selected by more than one Faculty member and then a F1000 Factor is worked out by means of a calculation that takes into account all the individual ratings received and the number of times a paper is selected by different Faculty members. Papers are also classified into five types – New finding, Technical advance, Interesting hypothesis, Important confirmation, or Controversial findings – and categorized according to all relevant subjects (creating a cross-linking between traditional disciplines). F1000 also has some interesting features, such as 'Hidden Jewels', which are articles picked from journals that are less widely read and so may have been missed by many researchers. F1000 Biology has been around since 2002 and is now well established. It was joined in 2006 by Faculty of 1000 Medicine (www.f1000medicine.com), which is based on the same underlying principles and functionality, and is collecting and organizing the opinions of researchers and clinicians to complement more data-driven approaches to evidence-based medicine.

(3) New models of peer review and variations on the traditional model

A number of journals are experimenting with new models of peer review and variations on the traditional model. Some are listed and described below (alphabetically) to give readers an idea of the sorts of things being tried.

(i) *Atmospheric Chemistry and Physics* (www.atmos-chem-phys.org)

This open-access journal, which was launched in 2001, has a two-stage process of editorial review that involves open public review and interactive discussion together with traditional review procedures. Manuscripts are first subjected to a quick 'pre-screening' by the editorial board and if they pass that they are posted on the journal's website as 'discussion papers' (in *Atmospheric Chemistry and Physics Discussions*). There is then an interactive public discussion period of 8 weeks. During this time, comments submitted by the reviewers chosen by the journal (who can, if they want, remain anonymous), comments posted by any other individuals (who must disclose their identity), and the authors' responses are published with the discussion paper. Comments are moderated by the editor assigned to the paper and any that are inappropriate are edited out. There then follows a stage of revision and traditional (non-public) review if required before publication of accepted papers in the journal proper. All versions of the article and all postings remain permanently with the article.

(ii) *Biology Direct* (www.biology-direct.com)

Biology Direct (an open-access journal published by BioMed Central) was launched in July 2006 and has adopted a radically new approach to peer review – it is the authors themselves who must find reviewers for their articles. Authors must approach the journal's editorial board direct and get three members either to review their article themselves or to obtain reviews from external reviewers. If they are unable to do this, the article is rejected. All the reviewers' comments, and their names, are published with papers. Even if reviewers express reservations about a paper, or challenge its conclusions, the authors can still elect to have their paper published unless there are ethical or scientific problems (for example it is 'pseudoscience'). Authors can, however, decide they do not want to have their paper published after review if they do not want to see it published alongside the reviews that have been received.

(iii) Cochrane Collaboration (www.cochrane.org)

The Cochrane Collaboration, which was established in 1993, is an international, not-for-profit, independent organization that provides up-to-date, accurate information about the effects of healthcare interventions, with the aim of helping people make decisions about healthcare. It regularly produces, updates and disseminates systematic reviews (Cochrane Reviews), developed from protocols, and has many thousands of contributors worldwide working collaboratively on this, often in teams. Both pre- and post-publication review are involved and it is an open process, with all contributions acknowledged. Rigorous quality standards are maintained and support and training are provided.

(iv) *Electronic Transactions on Artificial Intelligence* (*ETAI*; www.etaij.org)

ETAI is an open-access journal that was started in 1997 and uses a two-step, open peer-review model. On submission, articles that fall within the scope of the journal are posted on the *ETAI* website and the relevant peer community is alerted by email (this being possible because the fields are quite small). There then follows a discussion period of a few months during which all individuals who comment are identified and the author can interact with them and respond to their comments. The comments are screened to make sure they are relevant and not of unacceptable quality or content. The authors are given the opportunity to revise their manuscripts before they are sent out for external peer review. The reviewers' names at this step are not revealed and they are directed just to advise whether or not a manuscript should be accepted.

(v) *Journal of Interactive Media in Education* (*JIME*; www-jime.open.ac.uk)

JIME was launched in 1996 and has an open peer-review process that has three stages. Firstly, in a 'private open peer review' step, submitted articles are assessed by selected reviewers who post their reviews onto a private website; disclosure of their identity is encouraged but not required and they can email private comments to the editor. Authors can respond to these reviews and can enter into discussion with the reviewers for an agreed period of time. Reviewers are also encouraged to debate amongst themselves. Secondly, if the editor judges the submission to be of sufficient quality to be accepted, the paper is published as a pre-print for discussion, with the author–reviewer discussion, in a 'public open peer review' step. The relevant community is invited to enter the review debate, and the authors and reviewers may also respond. After a month, this review stage is closed and the editor posts an editorial report summarizing the most significant issues and specifies to the authors the revisions required, sending any private comments by email. Thirdly, after articles have been revised, they are published in final (freely accessible) form, along with those comments the editor decides are the most interesting. Further comments can then be posted by readers, and authors can post links to subsequent publications. Individuals can sign up to receive email alerts to new postings.

(vi) *Medical Journal of Australia* (*MJA*; www.mja.com.au)

The *MJA* has run two separate trials of open online peer review, the first in 1996 and the second in 1998, to explore whether different models of review could improve the quality and outcome of peer review. Extensive and detailed analyses of the trials were carried out. Readers wanting to find out more should go to the *MJA* website and to an article by C. M. Bingham *et al.* published in *The Lancet* ('The Medical Journal of Australia internet peer-review study', 1998, **352**, 441–445).

(vii) *Nature* (www.nature.com)

In June 2006, the journal *Nature* introduced an online peer-review trial which ran in parallel with its traditional peer-review process for a number of months (see www.nature.com/nature/peerreview/index.html for details; see also www.nature.com/nature/peerreview/faq.html for some very useful and thoughtful questions and answers that provide clarification on the trial details and cover possible concerns and implications for authors). All authors whose manuscripts were selected for external peer review were offered the opportunity to also have their manuscripts placed onto an open website (a pre-print server) where anyone, including the authors themselves, could post comments as long as they identified themselves and gave an institutional email address. All comments except those that were irrelevant or inappropriate were published. When the editors had received all reviews from the reviewers they had chosen (who remained anonymous), the open peer-review process closed, the manuscript was removed, and no more comments could be posted online. The editors then considered all the comments from both the private and the public review in their decision making. They also assessed the value of the public comments and measured the level of participation. The results and conclusions will be reported in *Nature*. The journal will also decide whether or not its traditional peer-review process should be changed in any way as a result of the trial.

(viii) *PLoS ONE* (www.plosone.org)

PLoS ONE is an online, open-access journal being launched in 2006 by the Public Library of Science (PLoS) that will publish research from all disciplines in science and medicine. It will use a review system that combines pre- and post-publication peer review. In the pre-publication stage, members of the editorial board (with input from external reviewers if required) will assess papers for publication on merely technical grounds, not subjective ones (so papers will be published as long as they are sound), and their comments will be available alongside the published papers. In the post-publication stage, readers will be able to post different sorts of feedback: 'annotations' will be short notes overlaid on the papers and are to be used to highlight a point, to make additions or corrections to specific points in the text, or to add web links to other material; 'debates' will contain the more conventional type of comments, responses and discussion; 'ratings' will be assigned to papers by users and then aggregated to provide overall ratings and so to rank papers. All contributors will have to identify themselves.

(4) Feedback from authors

An idea for improving the peer-review process was published in 2005 in a letter from Alon Korngreen ('Peer-review system could gain from author feedback', *Nature*, 2005, **438**, 282), where he suggested that authors should be given the opportunity to provide feedback (anonymously if desired) on the reviews they receive for their manuscripts. Journals could cross-reference feedback to the names of reviewers. This process would, over time, identify reviewers whom a number of authors had found to be deficient in some way. This is an interesting suggestion, and one that offers considerable scope for identifying reviewers who regularly fall short of required standards and should not really be used. Questions in the feedback forms could cover all aspects of review quality, accuracy and usefulness. It would probably be a good idea to also note the decision for manuscripts, as feedback may be harsher in cases where authors have had manuscripts rejected than when they have had them accepted, and to take this into account.

Index